J. T. Thirlborn 8/78
(SOC)
University of East Anglia

ECONOMICS OF MINERAL ENGINEERING

ECONOMICS OF MINERAL ENGINEERING

An inter-regional seminar organised by the United Nations
in co-operation with the Government of Turkey

Ankara, April 1976

MINING JOURNAL BOOKS LIMITED
LONDON 1976

Published in co-operation with the United Nations

First Published 1976

ISBN 0 900117 10 9

Text set in 10/12pt Monotype Times New Roman, reprinted photo-lithography by Chandlers (Printers) Ltd., Bexhill-on-Sea, East Sussex, England.

CONTENTS

OPENING ADDRESS TO THE SEMINAR

by Geoffrey Robson

Centre for Natural Resources, Energy and Transport

IT IS MY PRIVILEGE to join Mr. Shallon in welcoming you to this United Nations meeting on the economics of mineral engineering on behalf of the Secretary-General. This is one of a series of meetings that the United Nations is organizing in co-operation with interested governments—a meeting on mineral economics was organized in co-operation with the MTA Institute in 1970. A meeting on the Application of Advanced Mining Technology was held in Ottawa, Canada in May 1973, a meeting on the Negotiation and Drafting of Mining Development Agreements was held in Argentina in November, 1973, and a meeting on the subject of environmental factors as they affect mining industry operations is planned for 1978 to take place in one of the countries of Southeast Asia.

The association of the UN and Turkey in the form of technical co-operation between the Centre for Natural Resources, Energy and Transport and its predecessor, the Resources and Transport Division, and the MTA Institute of Turkey has been a long and fruitful one. It would not perhaps be inappropriate for me to explain that the MTA Institute has grown from very modest beginnings to become the present large and well-equipped organization. In 1935, the Mineral Research and Exploration Institute of Turkey, MTA, was established to explore the mineral resources of Turkey systematically and with responsibilities for the improvement of mines and quarries and for the training of engineers and technicians for the mining industry. During the coming two weeks participants, through visits to the Institute itself, and by participation in field excursions will be able to judge for themselves how well the Institute, under the distinguished leadership of Dr. Sadrettin Alpan, its Director-General, has discharged these responsibilities. That the MTA Institute is one of the largest of its kind in the world is in itself a tangible demonstration of the importance placed by the Government of Turkey on establishing a strong national organization to service mineral resources development in the country. Turkey's established reserves and untapped potential for a wide variety of minerals are large. The country is already established as a major exporter of a number of minerals, of chromite, borates and copper, among other mineral commodities.

The United Nations is playing a small, but, I think, effective role in assisting the Institute in its operational programme, notably by providing the services of certain key experts and in arranging for training of young Turkish professionals. At the present time Dr. Alpan and his senior staff are considering an expanded programme of UNDP technical co-operation in the 1977–81 period, which will include the establishment of a remote sensing data and interpretation centre which will serve not only the mineral sector but also be available for applications related to agriculture, forestry, water resources management and the environment. The setting up of a mineral data bank and of a research centre for investigating solar and other new energy sources which can find application in Turkey are all evidence of the dynamic approach being taken by the MTA in pursuing new techniques.

The willingness of the Government of Turkey to provide host facilities for this meeting demonstrates a concern for the minerals sector beyond the national context, and it may not be inappropriate for us at this meeting on the economics of mineral engineering to review the international background and the recent developments concerning economic relations between countries which are producers of minerals and other primary commodities and consuming countries.

The Sixth Special Session of the United Nations which took place in 1974 happened at a time when to the well-recognised differences in the level of economic development was added the perception on the part of many of the developing countries, that their rate of economic development compares unfavourably with that of many of the developed countries, that price trends for primary commodities were unfavourable in relation to those for manufacture and, therefore, their economic future seemed to them to be in jeopardy. From this Sixth Special Session emerged the call for a new economic order which for the developing countries would lead to an improvement in their terms of trade and a transfer of real resources to their benefit. The matters discussed at the Sixth Special Session were taken up again at the Seventh Special Session on Development and International Co-operation which was held in September, 1975. In contrast to its predecessor, that meeting was marked by a willingness to negotiate and avoid confrontation on the part of both the developed and the developing countries.

A conference on International Economic Co-operation began in Paris in December, 1975, and is now continuing there. These recent conferences on economic co-operation and development surely mark a growing appreciation in all countries of the real difficulties in which many of the developing countries now find themselves. There are probably few economists who would dispute that market forces in the past have tended to distribute benefits disproportionately so that increased productivity, whether primary or in manufacturing has tended to benefit the developed countries most.

The Paris talks and the recent European Common Market plan to stabilize developing countries' export revenues appear to be indications that a serious analysis of the problem, coupled with some remedial action, are in progress.

So far as the metalliferous ores are concerned the problems of stabilizing prices and modifying patterns of investment and revenue distribution are formidable ones, but since the economic well-being of many developing countries depends heavily on the mining sector, we must hope that the present international climate of negotiation and progress will lead to a solution to these problems. It is against this international background that we hold our meeting here in Ankara. Though our meeting is concerned with but a small segment of the issues I have referred to, we can hope that our work, which will include an analysis of the pattern of world trade, of the factors involved in financing mineral development, a review of new technology as it may effect mining development, together with an exchange of views and experience on the part of our lecturers, our participants and our hosts, will carry us forward in understanding each other and our perspectives and problems. You are after all, a group of people, both lecturers and participants, prominent in your own countries and working in the mining sector. Perhaps if we can maintain a willingness to understand the perspectives and needs of others and avoid the extremes of confrontation, this meeting may serve not only to assist mining development in developing countries but also to help towards a reconciliation of the larger issues.

PAPER 1

DEVELOPING COUNTRIES POSITION IN WORLD MINERAL TRADE

The changing pattern of processing facilities

by Michael West

Following an outline of the total value of world mineral and metal production, the particular contribution of the developing countries is given and the production and export patterns of the more important metals are discussed. This shows a mixed position of processing in both the consumer nations and the producer developing nations. The original smelters and refineries in the industrialised countries were established to process local production but as the mines declined, the countries became raw material importers and these imports have become an important element in their minerals pre-emption policies. Overall, the smelting/refining industries in the industrialised nations have declined in relation to total demand and the impact of changes, such as environmental programmes, is highlighted. Smelters and refineries in these countries have played an important role in developing new mines overseas.

Looking to the future, the assessment of integration projects is discussed and the need to complement engineering assessments with both funding and marketing studies is stressed. In this context, the inter-relationship of funding and marketing is described and the various funding and marketing arrangements associated with particular metals are contrasted. The competitive nature of mineral sales is outlined and the differences in the character of competition as between unrefined and refined materials is noted. Outline costs of integration projects for copper and aluminium are given, together with the possible foreign exchange benefits. In conclusion, there is mention of integration beyond the refined product stage.

AS AN ORDER of magnitude, the non-ferrous mineral and metal production of the world in 1973 was valued at $34 billion*. In 1974 the value of mineral production was sharply higher at $48 billion reflecting the peak prices for a number of products and, for 1975, preliminary estimates suggest a total value somewhere between these figures. It would not be appropriate to base general considerations on the more recent figures as 1974 and 1975 included marked anomalies both as to price and apparent consumption. To the 1973 total must be added the value of iron output which, on an ore basis, would be a further $5 billion or, assessed as pig iron, $20 billion,

* This figure excludes the production of the centrally planned economies, see appendix for list of minerals and metals included. NB: 1 billion = 1×10^9.

The Developing Countries Contribution

The total of non-ferrous production may be divided into three distinct categories—the value of mineral production arising from output in the developing countries, the value of mineral production from the industrialised countries* and the value of secondary production in the industrialised countries. (Secondary output within the developing countries is negligible.) The respective figures are $13 billion, $17 billion and $4 billion, or, expressed another way, 38%, 50% and 12%. If the United States' contribution is excluded from these figures, they reflect more realistically the position of the developing countries' contribution to the international minerals trade. The total value of non-ferrous production, excluding US, is some $28 billion of which 46% is contributed by the developing countries and 7% represents secondary recovery in the developed nations. Broadly, these figures have been assessed on the basis of the value of refined production but, with the wide variety of minerals traded and the differences that exist in recognised pricing points, these totals can only be regarded as 'order of magnitude' figures.

On this basis, the value of non-ferrous mineral resources represents about 1% of the world's Gross National Product—or, divided proportionately, some 0.7% of the GNP of the industrialised nations and 3% of the GNP of the developing countries. It can be argued that this is an unreasonably low value to place upon the mineral raw material input into an industrialised society, and there seems little doubt that, over the post-war years, the proportionate contribution of mineral raw minerals has declined. Although this particular problem is not directly relevant to the question of improving revenue through further beneficiation, it remains a basic market factor and will intrude again later. Meanwhile, it should be noted that for most mineral products there is an internationally accepted price and that the revenue received per refined unit varies little at a point in time. There is no discrimination against the less developed countries in this context; indeed, to the extent that alternative price mechanisms exist, they can work against the primary producer in an industrialised country.

Direct consumption of locally produced minerals in the developing countries is very low and thus the overall figures suggest that there is a transfer of raw material resources valued at $13 billion from the developing countries to the major consuming nations. This valuation based on the refined product price, however, overstates the actual revenue. In many instances, the mineral resources are shipped in an unrefined form and thus the actual revenue received by the exporters will have been lower. Again, as an order of magnitude the 1973 value of minerals and metals, as exported, was about $9 billion or 70% of the refined value.

These overall figures mask the significant differences that exist between the individual mineral products. Copper is outstandingly the single most

* The OECD nations together with South Africa.

important item, representing almost 45% of the total. Aluminium is second but, with a relatively low-level of local benefication, the income to the developing countries is low and its contribution to the total is only 11%. It is closely followed by tin at 10% and diamonds are the only other single product representing more than 5% of the total.

During 1973, the developing countries contributed almost 50% of world primary copper production and 73% of this total was exported in metal form (21% as blister or unrefined metal and 52% as refined shapes). The developing countries' contribution to world bauxite production was 56% but only 14% was exported as metal ingot. The export patterns of copper and aluminium may be further contrasted in that virtually no copper is exported as ore and only 27% as concentrates; on the basis of contained aluminium, 56% is exported as bauxite and 30% as alumina. Even these figures overstate the aluminium position slightly as there is some smelting in the developing countries which does not arise from domestic bauxite, or alumina production.

The iron situation is not directly comparable with most non-ferrous metals as a number of developing countries are steel producers and, in total, contribute about 6% of raw steel output. This production, however, is mainly to meet domestic requirements and is frequently based on local scrap. This steel production cannot be said to represent an early stage of integration for major iron ore producers.

Although there is a significant international trade in iron ore, much of it is within the industrialised group of nations; the developing countries contribute about 33% of the ore consumption. However, as there is a major scrap feed, the developing nations' contribution to total iron consumption is rather less than 20%.

In the steel making industry, two factors are dominant; firstly, iron ore output from the OECD nations, together with scrap arisings, accounts for over 80% of raw material feed in the industrialised nations. Secondly, there is no truly comparable ingot metal stage at which the product is transportable or can be said to be fungible—as most base metals. An interesting technological development which could lead to a situation more parallel to the base metals is the introduction of an intermediate stage, namely the metallised or pre-reduced pellet. Present indications are that the price differentials would not justify major developments in this field and a direct move to raw steel production is perhaps more likely. Meanwhile, it must be recognised that the steel works continues to be regarded as the first stage of a manufacturing industry, rather than a natural extension of a primary iron ore producer.

Smelting in the Industrialised Countries

The dependence of the European nations upon the developing countries for mineral raw materials began at the turn of the twentieth century. In part, this reflected the growing consumer requirements of the industrialised nations but, perhaps more importantly, it developed following the gradual decline of their own domestic mining industries. Until that time, Europe had been largely self sufficient and Britain, for example, was a major exporter of base metals through the nineteenth century. Thus, smelters and refineries were originally sited in countries that provided both the mines and the markets, though, it is interesting to note that, even at that time, the mines were dissatisfied with the location of the smelters which was dominated by the availability of coal rather than the concentrates themselves. It can also be noted that many of the 'developing countries' supplying Europe in the early 1900's are now, themselves, major consumers and this pattern of development can be expected to continue.

Without exception, mine lives are limited by the ore reserve position. This is not the case with smelters, however; they are subject to technological obsolescence but this is unlikely to coincide in any useful way with mine life, and furthermore, plant modification can and should be a continuing process. Thus, a smelting industry adjacent to a major market or with good shipping facilities can be sustained long after the original mining industry which supported it has declined. This natural sequence, as much as anything else, accounted for Europe's continuing role in smelting and refining. The close proximity of the market is a critical factor, for there is not only the convenience of maintaining a relatively limited stock of the wide variety of shapes required by fabricators, there is also the availability of secondary material. In practice, however, the position of the European smelting industry (aluminium aside) has declined relative to the massive growth in consumption.

In Japan, the position is rather different. Although the original smelters were established to handle domestic mine production, there has been a conscious policy to expand smelter capacity in line with consumption. Individually, the Japanese mining companies did not have the stature to develop new resources overseas to meet expanding demand. They could only maintain their position domestically, as suppliers of refined metals, by expanding their smelters/refineries and importing concentrates. More importantly, however, these concentrate imports formed part of a deliberate national minerals pre-emption policy; for a number of reasons, not least the over-supply position, this policy is now being reconsidered and modified. Nevertheless, the minerals pre-emption aspect of unrefined mineral imports is important.

In the aluminium field, the background of history does not intrude as with the older metals. Substantially, the siting of aluminium smelters has been dictated by the availability of abundant (cheap) electric power. Hitherto, this has meant a close link between hydro-electric resources and aluminium smelting. With improved electrical efficiencies, the power question is less dominant but even with low power costs, power consumption still represents 25% of smelting costs.

The other aspect which sets aluminium apart from the older base metals is the actual cost of refining. Taking copper as an example, the cost of copper in concentrates might represent about 70% of the value of a refined wire-bar. The cost of aluminium in alumina only represents some 40% of the final product and there is the further difference that bauxite is frequently the exported product (copper is rarely exported as ore) and the aluminium content of bauxite only represents 10% of the cost of a metal ingot.

Environmental Considerations

Over the past few years, the pre-occupation with environmental factors has deterred further smelter expansions in the highly industrialised centres. For a long period, there was a general excess of smelting capacity and importing smelters, operating on fully depreciated plant, were willing to offer attractive contracts to maintain adequate supplies. The position has now changed, however, the environmental restrictions have obliged smelters to reduce their throughput rates and, also require considerable capital expenditure to overcome emission problems. Most of these factors have served to increase charges.

In the short term, because of the general downturn in economic activity, the restrictions placed on smelting operations have not caused a general shortage of capacity, but this must be expected. As consumer requirements are re-established, new smelting facilities must be developed and this will demand a re-assessment of the general criteria. It can be assumed that in the industrialised centres the new environmental requirements will have increased the cost, in real terms, of smelting and refining. In periods of over-supply, this will inevitably restrict the price payable for unrefined raw materials, but, over the longer term, these added costs will be reflected in the price of the refined product. This would increase the revenue available to mine operators if they integrate through to the refined product at mine site. Clearly, the environmental programmes must also have their impact in the primary producing areas, but the density of urban and industrial development in most cases is such that a presently acceptable level of control can be achieved through 'dilution' of effluent rather than the total elimination of undesirable products. Furthermore, the cost of effective control in new plant is considerably less.

Detailed Project Assessment

The viability of any particular project can only be assessed by close reference to the detail of each situation. It will vary significantly from one product to another and also from one location to another. Labour, power and material costs will all be important factors as would be the cost of transport. Size and potential mine life will be another critical consideration as, in almost all cases, the more sophisticated benefication and refining techniques are capital intensive and there is a minimum viable size.

Engineering feasibility studies based upon such knowledge will give viability determinations that may be expressed in terms of profit—in the conventional domestic currency sense—or may be expressed in terms of foreign exchange benefit. It must be recognised that the latter is, on occasion, a more relevant concept for a state enterprise. In most circumstances, however, the detailed feasibility study is unlikely to show a markedly different cost position from that of a possible competitor in an industrialised area and, in these circumstances, other factors such as marketing and funding become significant.

Availability of Funds

Except to the extent that funds are available in the form of aid finance, marketing and funding are significantly inter-related, and this has been highlighted by the number of raw material contracts that have been used as collateral for loan finance over the past ten years. This is a field that has developed significantly and an examination of possible future developments is relevant.

The original concept of the minerals supply contract replacing a tangible asset as collateral for new mine financing grew out of the early post-war government purchasing contracts for uranium. Banks accepted these contracts as security and new mines were developed with a much larger debt element than had been usual hitherto. These proved successful and, at a second phase, lenders were prepared to advance funds on the basis of contracts with major private-sector companies. Typically, these were the smelters and refineries in the industrialised countries—especially Japan and Germany.

The fact that the 'consumer' was represented mainly by the smelter/refinery rather than the semi-fabricator is significant. In most countries, the semi-fabricating industry is too fragmented for any one unit to have sufficient buying capability to establish an adequate purchasing contract. Thus, the smelter/refinery was the natural focus point for the purchasing power. Secondly, semi-fabricators are not constrained to remain with a particular material in the same way as a smelter/refinery and contracts beyond one year have been unusual. The comparative uniformity of the refined product has meant that the semi-fabricator is less dependent upon a particular supplier; in a last resort, any refined product purchased on the open market would suffice.

For the smelter, and to a lesser extent, the refiner, the need to maintain stable supply patterns has always been important and the long-term contract was a familiar instrument. Furthermore, in Japan, the long-term contract associated with loan funds was encouraged by the national government as an expression of its raw materials pre-emption policy.

Over the years, these supply contracts were developed and modified to reduce the risk to lenders and, in turn, the cost of borrowing. This included the introduction of take-or-pay clauses and minimum price terms—both of which were designed to assure minimum cash flow levels to the new mining project.

More recently, the marked down-turn in demand for raw materials has introduced a new element as these apparently 'fixed' contracts have become 're-negotiable'. Not only have price modifications been sought but buyers have been forced to re-negotiate the tonnages agreed. This is clearly a development which is damaging to the concept of debt financing and will inevitably limit the lenders' willingness to support new mine projects in this way. It will reduce the amount of capital that is likely to be available for any given project and will, therefore, restrict an operation's ability to integrate beyond the first marketable stage. It may also mean that financing institutions will insist on lending to the customer (refinery or semi-fabricator which in turn will finance the mine) rather than lending to the mine operation directly. This pattern will obviously change as the availability of finance and the demand for metals improves but this is likely to take some years.

To place these loan funds in context, an integrated copper smelter/refinery project (including acid plant) for 60,000 tons per year would cost about $100 million. An integrated aluminium project, including both an alumina plant and a primary aluminium smelter to produce 100,000 tons of ingot metal each year would cost $250 million.

Given that most debt finance is available only for a maximum term of years, the inclusion of an integrated plant into a mine development project has a double impact upon the total borrowed. There is the direct additional cost of the plant itself. It may also be necessary, however, to increase the overall size of the project to assure an adequate cash flow to meet the additional borrowing for the integrated plant. In turn, the larger mine project itself will need additional funds and, in this way, total costs can escalate rapidly. There have already been examples of exactly this situation where the spiralling of capital costs seriously delayed a major project with the result that an opportunity to meet a particular supply deficit was lost.

An insistence by the mineral rights owner (frequently the local or national government) that an integrated project be established at the initial stage can be counterproductive. Examining the earlier mine developments in Europe and North America, the natural pattern was for the mine to be established initially with the earliest sales product possible, and only when the operation was self-financing, did integration proceed through to the refined metal stage. It will be desirable—even necessary in some cases—to incorporate into a mining licence, a commitment to further integration once the project is self-financing, but early initial development must be the primary aim. The agreements associated with the development of the Western Australian iron ore resources provide very good examples of commitments to further integration.

The agreement signed between Hamersley Iron and the Government of Western Australia, provided for an initial expenditure of $60 million in three years from March, 1965 to provide facilities for iron ore mining and export at an annual rate of 1 million tons. Hamersley Iron was further required to submit proposals within 10 years for a secondary processing plant with a capacity of 2 million tons a year and to submit proposals for an integrated iron and steel industry in Western Australia by mid-1986. The agreement and the company's Lease and other rights, thereunder, are subject to compliance with these obligations though it is accepted that the plant must be economically viable. Similarly, the Mt. Newman agreement sets out the company's obligations with respect to the development of both secondary processing facilities and the construction of a steel plant. It calls for the development of secondary processing at a rate of 500,000 tons per year by 1981 expanding to 2 million tons per year by 1985 and this is to be followed by steel or pig iron production at an annual rate of 500,000 tons by 1994 expanding to 1 million tons by 1999.

Competition and Pricing

Before turning to contrast the marketing of unrefined and refined products, it is appropriate to examine the more general questions of competition between raw materials and also of pricing. The comment is sometimes made that producing countries have little or no control over commodity price fluctuations. This is clearly true, but it must be emphasised that no other party—neither consumer nor intermediate market agent—has any control either.

There are important distinctions between the pricing systems for each of the major metals and these are reflected in the degree of price fluctuation that occurs over a given period. However, the ultimate price determinant, in every case, is the balance of production capability and real consumption. In the shorter term, the impact of production capability will be modified by actual availability; similarly, real consumption is frequently overshadowed by a marked change in demand reflecting the semi-fabricator or manufacturer's anxiety to increase or decrease stock levels.

The change in consumer stock levels is probably the most significant cause of short-term price fluctuations, and the greater the possible stock level shift, the greater the price movement. However, the underlying price level is determined by the basic level of productive capacity. If new producers enter the market prematurely (ie. before the growth in consumption requires the extra capacity), then prices will fall to, or remain at, generally low levels. This will encourage demand from consumers who might otherwise opt for an alternative material. (With the modern development of materials technology, there remain few applications where any particular metal can be said to be irreplacable.)

Clearly, if there is an excess production capacity at a point in time, then producers (acting jointly or severally) can modify availability by cutting back supplies. Experience with various commodities has shown that this can be effective when excess capacity arises from a down-turn in consumer demand. However, it is not effective when the excess capacity arises due to a new producer forcing his position into the market prematurely. This places an unreasonable burden on any formal or informal 'spirit of co-operation'. Throughout the history of commodity agreements, it has been accepted that new producers must establish their position for a period of years before they join the group and benefit from the collective support of cut-backs in periods of over-supply.

Although it is not normally practicable to distinguish the individual elements, new mine capacity year-by-year is of two parts. Firstly, there is the requirement to replace exhausted orebodies and, secondly, there is the programme to provide additional production to meet expanding consumer demands. Through the 1960's, it appeared that some 55% of capital expenditure on new operations and expansions was for 'replacement' facilities while the balance provided for real growth. To the extent that the new mine capacity represents real growth, there must be a parallel expansion at the smelter/refinery stage, but where new mines are replacing exhausted operations, it will not generally call for a changed smelting capability. In practice, it is not possible to determine those particular mines which represent growth, and others which represent replacement; nevertheless, it is clear that not every new mine can develop a new smelter unless other smelters are to close prematurely.

Mines and process plant are, to an extent, independent in that there have been periods when mine capacity was in surplus and others when smelter capacity was in surplus. However, there is an important inter-dependence in that additional smelter capacity in excess of requirements may not only create surplus in its immediate field, it can also prompt surplus capacity at the mine stage. A smelter or refinery in a consumer country with established customers, will normally itself stimulate a new mine development to replace a lost concentrate feed (following a mine site integration programme), whether or not the total market for the product justifies the additional mine. Because of the general growth in metals consumption over the past three decades, this situation has rarely been obvious, but, given a fixed consumption pattern it can be seen that an integration programme to treat the concentrates of an existing mine may stimulate the development of a new mine—probably in another country—to feed the original smelter. The result is not merely excess smelter capacity, but excess mine capacity; and it is the new mine that has direct access to the established group of customers through the original smelter.

Thus, it can be seen that the mineral producer is in a highly competitive position and that competition will arise from two main areas; from other potential mines (frequently in other countries) seeking to develop reserves of the same metal and also from the producers of other metals. Competition from other metals again emphasises the new freedom of choice given to the minerals consumer by modern materials technology.

Marketing

The counterpart to minerals pre-emption by the industrialised countries is market pre-emption by the primary producers. Hitherto, this complementary programme has taken the form of either corporate integration across national boundaries or long term contracts—or, of course, a combination of both. Looking to the future, the need to secure access to a particular market sector will be as important as ever but with increasing upgrading of mineral products in the producing areas the 'consumer' will tend to be further along the line of integration. For example, it will be the refiner rather than the smelter, or the semi-fabricator rather than the refiner, who will negotiate the agreement with a primary supplier.

To some extent this pattern is already developing naturally in both producer and consumer countries but the initiative now rests increasingly with the primary producer. Until recently, the intermediate processors in the consumer nations have had adequate purchasing power within single corporate units to stimulate the development of new mine capacity. As this is rarely true at the next stage of vertical integration, the would-be mine site smelter/refinery must take the initiative and propose collective buying by the semi-fabricators and even propose advantageous buying and stockholding arrangements.

There is, inevitably, a cost to the provision of such facilities and this is frequently under-estimated in the conventional engineering assessment of the viability of local treatment programmes. The foreign exchange revenue from ore or concentrate sales is often available once shipping documentation is complete. If a refined product is to be sold to semi-fabricators, then the revenue may be delayed through to delivery and this will not only include shipment time, it may also include a significant stock holding period if a competitive service is to be provided by way of variety and shapes. (This situation varies from product to product.)

Service of this type is as important in the competitive sense as between one metal product and another as it is between suppliers of similar products. It is arguable that the competitive element between metals is an industry-wide matter but it is significant that materials like aluminium, which are mainly 'produced' in the consumer countries, have enjoyed the strongest growth over the past two decades. This primarily arises from considerations of stability of supply on the one hand and, most importantly, from far more vigorous marketing programmes by the consumer-nation based producers on the other. Outside of the centrally planned economies, very little of this can be ascribed to preference for domestic goods in the foreign exchange saving context.

While there are other contributing factors, it is very significant that a growth-rate ranking of the major base metals runs almost exactly parallel to a ranking reflecting the proportion of world 'production' contributed by the industrialised nations. Aluminium and nickel have shown exceptional growth while tin, with hardly any direct support from consumer-nation based producers, has shown least growth. Both copper and zinc have shown moderate growth but again, with its greater industrial nation output, zinc has done better than copper. Lead is the only anomaly in this situation but it has suffered from severe environmental restraints.

This evidence suggests that similar factors are also at work between individual mine operators within one metal commodity. However, the development and decline of individual mines, or even national industries, is dominated by ore reserve life and the parallel of market development cannot be demonstrated.

Funding Possibilities

Given that an integrated project is viable in the accepted commercial sense—and this does not just mean that there is a net foreign exchange benefit—then the question of capital availability arises; with a satisfactory sales contract arranged, then there should be little difficulty in arranging funds. A smelter or refinery project is very different from a new mine in that the lead time is shorter, the risk element is lower and, importantly, the plant content of the total cost is normally higher.

Aside from aid funds that might be available for associated infrastructure from supra-national institutions like the World Bank or UNDP, the major sources of finance available are the manufacturer credits (supported by the various export credit guarantee schemes) and the international banking sector through the Euro-currency market. Maximum use must be made of the manufacturer credit schemes as the terms are normally more attractive than the commercial money markets. These schemes frequently extend to include the design and consultancy costs and, if there is a problem, it is that the variety and apparent attractiveness of the financing schemes can overshadow the intrinsic merits of the various plants offered.

Another element of cost which is funded relatively easily (though not necessarily cheaply) is the transit pipeline stock as the refined metal itself is a satisfactory collateral. However, at least a portion of the production will already be committed to the contract buyer as collateral—directly or indirectly—and this will add to the complexity.

The arrangement of the marketing contract is the most difficult. As emphasised earlier, the main sales contracts hitherto have been with smelters and refineries and it is likely now that it will be necessary to establish new ground by arranging sales contracts with semi-fabricator groups.

Although the London Metal Exchange is frequently criticised as a market prone to exaggerate price movements (a reasonable criticism of the system, though perhaps not of the LME as such), it does have the value of assuring all producers with a similar unit revenue and of providing—through its contango mechanism—a buyer of last resort. This is an important factor and one which can be particularly significant to the smelter established in a mineral producing area independent of any major consuming interest.

In contrast, the so-termed producer price mechanisms offer greater stability. However, in times of oversupply, discounts develop which are far from uniform and an independent producer, without an established market outlet, can be at a major disadvantage. Furthermore, with the producer price systems, there are rarely any buyers of last resort who are in any way comparable to the financial institutions which hold terminal market stocks.

Although a much longer-term development, there is pressure for greater commodity price stability and metal stocks are likely to be part of any stabilisation scheme, through one mechanism or another. However these develop, it seems most likely that access to these stockpiles would be available only to those able to deliver refined metal, probably in ingot form. The possibility of toll refining would assure that the non-integrated producer was not entirely excluded but this possible development would encourage integration.

Although toll smelting and refining in a consumer country has an application in this context, it would not help particularly in the marketing context. An alternative to mine site refining is for the mineral producer to participate directly in a refinery based within a consumer area. This can be achieved by linking the value of the raw material to the final product—as has been done with bauxite exports from Jamaica—but this penalises the particular exporter, in the competitive sense, against alternative suppliers.

A better approach is to secure a participation in the effective equity of the refinery as, for example, with Australian zinc concentrate exports to Europe. This not only secures access to a market, it can also enhance foreign exchange revenue by participating in profits—including profits arising from by-products often not saleable outside a major consuming area. Incidentally, this type of participation can also provide a technological involvement which would be invaluable when further expansion suggests that mine site refining is appropriate.

The net additional foreign exchange revenue that may be earned by producer area processing varies significantly from product to product but the added value is most often a function of the cost of capital plant rather than the local labour input. In almost all cases of integration beyond the concentrate stage, the process is highly capital intensive. The capital cost of a large integrated copper plant from concentrates to wirebar is of the order of $1,500–$1,750 per annual ton. An integrated aluminium plant, from bauxite to metal ingot will cost of the order of $2,500–$3,000 per annual ton. Amortised over 10 years at, say, 7% and allowing for a two year development period this gives a cost of $265 per annual ton of copper and $450 per annual ton of aluminium, representing about 65% of the likely added foreign exchange value in each case. The local labour costs in both copper and aluminium are, also, remarkably similar, but the materials and power costs are greater in aluminium production (a factor which has led to the development of aluminium smelters away from either bauxite producing areas or even major market areas).

The basic decisions on whether or not integration can be justified in foreign exchange benefit or conventional profit terms can only be assessed on case by case studies. These will involve, primarily, assessments of the likely life of local resources, the availability of other necessary raw materials, the cost of power and the relative transport costs. The assessment must also accept that capital funds are limited and contrast such an integration project with other opportunities—possibly not even in the minerals field. Given that the assessment indicates a positive advantage, then the question is one of timing and it is at this stage that marketing studies become critical. In certain cases, integration is technologically unavoidable as an initial programme, but where it is not essential it is rarely desirable and is best scheduled as a second phase.

Looking to the longer-term future, it is likely that most of the net growth in smelting and refining will take place in producer countries but, in parallel, the consumer nations are likely to strengthen their domestic capability as far as possible by improving secondary metal recovery and by mining alternative materials. None of this, however, will modify the fundamental inter-dependence of producer and consumer. From time to time, surpluses and shortages do shift the balance of strength from one to the other and the effect is real, whatever the particular cause of imbalance may appear to be. However, an increase in the level of integration by the primary producer does not fundamentally modify the balance of inter-dependence.

Integration beyond the refined metal stage becomes progressively more difficult for producers isolated from a major market; equally, the added value of the fabricated product is significantly greater. In the United States, for example, the value of both refined copper and aluminium is trebled in the form of processed materials. This will include the manufacture or some relatively sophisticated products, but it does indicate that once some refined production has been established it can be more valuable to move towards a semi-fabricated production for a portion of output rather than to seek total integration at the refined metal stage.

Discussion Summary

The discussion which followed highlighted problems and policies of individual developing countries including the possibility of backward integration, from a rolling mill to an iron and steel plant and then an iron ore mine, for example. The need for earning and saving foreign exchange, the improvement of infrastructure, technological spin-off and industrialisation by tertiary development of part of production were reviewed, as was the desirability of agreement among producers on market-based share of increases in production, ensuring harmony by setting limits on foreign exchange flows by consensus and agreement with consuming countries as in the case of the Lomé Convention.

REFERENCES

International Monetary Fund, Annual Report, 1975.
Metal Statistics, 1964–74, Metallgesellschaft AG.
Mining and Minerals Policy, 1975. Annual Report of the Secretary of the Interior, USA, under the Mining and Minerals Policy Act, 1970.
Mining Annual Review, 1974 and 1975.
Copper Task Force Report, British Columbia.
Aluminium Industry Report, Oppenheimer and Co. Inc.
Statistics of the World Trade in Steel, United Nations.
American Bureau of Metal Statistics, Year Books.
Capital Requirements of the World-Wide Mining Industry, M. West, **American Mining Congress,** 1971.
Iron Ore (Hamersley Range) Agreement Act 1963 (No. 24) Government of Western Australia.
Iron Ore (Mt. Newman) Agreement Act 1964, Government of Western Australia.

APPENDIX

The following metals and minerals are included in the survey: copper, lead, zinc, tin, bauxite, gold, diamonds, platinum, manganese, nickel, uranium, chrome, mercury, molybdenum, asbestos, boron, potash, phosphates, ilmenite, rutile and zircon. These metals and minerals account for some 94 per cent of non-ferrous mining activity. For further details see "Mining Activity" **Mining Magazine,** September, 1968, page 217.

PAPER 2

THE ROLE OF MINERAL ENGINEERING IN RESOURCES DEVELOPMENT IN DEVELOPING COUNTRIES

by Henry E. Cohen

Mineral Resources in Developing Countries

ALL COUNTRIES OF the world are developing, sooner or later, faster or slower, forward or backward. They cannot stand still because they are subject to a wide range of changing internal factors and external circumstances which cannot remain static. However, the term "developing countries" is now employed in a more restricted sense to describe those that have lagged behind the average level of material prosperity attained by other countries that are regarded as advanced. The reasons for the lag as well as the criteria of advanced status are manifold and are rarely capable of simple definition. Advanced status has been often correlated with possession of mineral resources, but the correlation is neither simple nor universal. Switzerland is an advanced country, yet its mineral resources could be regarded as negligible compared with those of Greece, Spain or Portugal. These three countries are less advanced at present although each was accepted as very advanced at some period in man's history. Denmark has far fewer mineral resources than Norway, yet the two countries are equally advanced. There are great numbers of combinations of contributing factors which control a country's state of development. Vast influence is attached to local circumstances, to human and to social factors which are quite outside the scope of this paper. The possession of mineral resources is itself greatly dependent on chance factors of geography and geology, but proper utilisation of mineral resources can offer fundamental advantages of development potential, which few countries can afford to ignore.

In principle, every country has four basic resource commodities on which its potential for prosperity must depend: these are people, agriculture, mineral deposits and geographic position. The last two are key factors in any national progression from a level of agricultural subsistence to a prosperous industrial and trading society. Accidents of geography and climate control a country's potential for agriculture and for trade. Similarly, accidents of geology and structure of the earth's crust control a country's accessible mineral resources. Indeed, the irregular distribution of resources makes trading a universal necessity and the prosperity of any country depends ultimately on the development of adequate trading commodities which are in international demand. In this context the export of mineral raw materials is unsatisfactory as a basis for national development. The contained fob values of direct shipping ores or of concentrates are generally quite low, ranging from around \$15 per ton for direct shipping iron ores to about \$3000 per ton for high-grade tin concentrates. In most cases, it is obvious that the added value of the refined metal or the manufactured product after smelting and primary processing is very considerably larger and that the gain greatly exceeds the additional costs. More important, the overall benefits to national prosperity derivable from mineral resources are vastly enhanced by the complex developments of infrastructure, engineering services, public utilities, transport and service industries which are inevitably associated with processing beyond the simple extraction of ores.

The most important consideration, however, is that the development of refining, fabricating, manufacturing and service industries produces local investment in education as well as diversification of local employment. This last factor is the vital link which alone ensures the permanence of prosperity. Few mines have productive lives exceeding some 30 to 50 years. They truly are wasting assets and offer no security of long-term national development. The most apposite form of utilisation of these assets is to convert them into skilled manpower to form a more permanent and adaptable asset for long term deployment after the primary mineral resource has become depleted. A logical sequence thus becomes discernible for developing countries:

(i) Their natural mineral resources should be regarded as bases for regional development rather than simply as assets for commercial exploitation. Even if mineral production were immediately commercially viable in its own right, it should nevertheless be deployed as part of a much broader plan. In the first instance it should yield a maximum incentive for the local development of its own necessary support facilities, including skilled manpower, engineering services, communications, administration, etc. This should be undertaken as an exercise in planned transfer of technology with an investment in local education and training far greater than would be justifiable by the commercial capitalisation of the initial mineral asset. It should aim at producing the capability for progressive expansion of local processing of the ore, from initially simple production of concentrates to the highest level of added value processing sustainable by local competence. This evolutionary development may be expected to occupy the best part of 20 years from first inception of a project. During the next phase this should lead to industrialisation based on the mineral resource.

(ii) The local skills and facilities built up during the first period should form a natural basis for a second phase of associated wider industrialisation and development of local resources, including industrial farming and more diversified exploitation of mineral deposits. New industrial developments should be designed to become at least self-supporting within their periods of capital life. In preparation for the phasing out of the original mineral resource, this second phase of development should be essentially one of expanding and transferring the facilities and skills that were developed locally.

This extended two-stage evolutionary development attempts to take proper account of the fact that complex industrial technology is indivisible.

Selected parts of technology may be transferred for the sake of speedy development of a mine and on an assumption of sustained back-up support from developed outside sources. However, this tends to lead to a self-limiting type of satellite activity which extracts the mineral resource but fails to convert the asset into local industrial competence.

The Nature of Mineral Resources

Mineral resources comprise all types of rocks which can be made to yield useful metals or industrial minerals. Under normal commercial circumstances two constraints need to be added to the above statement:

(i) It must be technically feasible to extract the useful products at costs within the limits imposed by their commercial value;

(ii) The size of a deposit and its value content must be large enough to justify the capital investment needed for its exploitation as well as to sustain associated developments and commercial commitments for an appropriate period.

For example, conventional integrated iron and steel production using blast furnaces and LD converters is unlikely to be viable unless a supply of high-grade ore or concentrate of at least 6 million tons per year can be guaranteed for, say, 25 years. If a low grade magnetite deposit were available, assaying 30% Fe, this might be turned into a concentrate assaying 60% Fe. Since iron recovery would be less than 100%, more than 2 tons of ore would be needed per ton of concentrate. Thus 6 million tons of concentrate per year would need about 15 million tons of ore. For a 25 year period this would necessitate a minimum of 375 million tons of reserves. In contrast, the basic requirements of a mini steel works project might well be met from reserves of 150 million tons, provided an adequate grade of concentrate is attainable.

From the point of view of exploitation it is convenient to divide mineral resources in general into three basic types, irrespective of their geological characteristics:

(i) ores containing a commodity of high price in relatively low concentration, rarely exceeding 1 or 2% by weight. Well known examples include ores of precious metals, uranium, nickel, tin, vanadium, diamonds, and many others. It may be stated with conviction that these commodities are rarely found in concentrations that would give commercial quantities of the ore a higher contained value than about $60 per ton at present prices. The lower limit or cut-off grade is usually defined by the specific mining and processing costs, but is generally not below about $30 per ton at present;

(ii) ores containing a commodity of low price in relatively high concentration, in the range of 20% to 90% by weight. Common examples include ores of iron, manganese, chromite, aluminium, coal, gypsum, fluorspar, and many more. None of these commodities occur with 100% purity in nature, but many are technically pure enough to fall within user specifications. These are termed direct shipping ores and require a minimum of preparation for subsequent processing. The cut-off grade for low grade ores which require concentration to become acceptable is again determined purely by the balance between the costs of mining/processing and the market price. Transport costs play a major role and often account for half the total costs with a bulky commodity of low unit price;

(iii) as an extension of (i) and (ii) some commodities have attained a rather special position due to remarkable combinations of circumstances: they occur in low concentrations in ore bodies of exceptionally large size and situated near the earth's surface. This facilitates very simple open pit mining on extremely large scales, up to 100,000 tons per day or more and thus permits reducing the costs per ton through economies of scale. If the same principle is applied successfully to concentration, commercial viability may be established if the market can absorb the resultant large quantity of product. Two outstanding examples are low-grade ores of copper and of iron. Huge porphyry or volcanic orebodies containing 1% copper are successfully exploited by cheap mining in very large open pits, followed by concentration employing efficient, highly-selective froth flotation. The contained value of such an ore is only about $12 per ton at a current copper price of $1,200/ton. However, it must be admitted that commercial survival is often facilitated by the presence of precious metal by-products (usually gold or silver) which can almost double the contained value of the ore.

The other example in this class is provided by low-grade iron ores. These are found in bodies containing many hundreds of millions of tons at grades between 20 and 30% Fe. They are mined openpit at rates up to 30 million tons per year or more, concentrated to around 65% Fe, pelletised and shipped all over the world with a market value of around $25 to $30 per ton of pellets. At that price the contained value of the ore is only about $10 per ton and there are no hidden by-products or other subsidies.

The three types of mineral resources indicated above present different problems of exploitation, of development and of integration in regional development projects. These various aspects are discussed below, but it should be noted that few countries have a free choice of the type of deposit they may wish to develop. It is generally prudent to assume that in many regions ores that are easily accessible and of profitable commercial quality have been found and exploited by traditional private or public organisations. The ore deposits that remain are often too inaccessible, too low in grade, too difficult to treat or too small in size for normal commercial profitability. It is therefore necessary not only to identify any such mineral resources with reference to the three basic types listed above, but to assess whether they offer potential as bases for regional development.

The investment required for opening up a mineral resource may or may not become recoverable indirectly during second stage industrialization, but it is desirable that mineral production is at least able to produce income towards the development of its own back-up facilities. It is generally true that a small mine reaches production and revenue faster than a large mine and the same applies to openpit working in comparison with underground mining. This points towards a preference in principle for seeking a pilot scheme of modest size, based on an accessible higher grade part of any large, low-grade ore body. The pilot project might gradually expand to support sound regional development even if conventional commercial profitability is demonstrably unattainable.

The Role of Mineral Processing

In any project of exploitation of mineral resources, the primary mineral processing, in its many diverse forms, plays an important role which needs to be defined clearly in relation to the aims of the project. The terms of reference can be chosen arbitrarily from a number of alternatives to meet specific purposes. For example, it may be desired to extract the maximum amount of a commodity from a limited source; a raw material of low value may have to be produced from large reserves at the lowest possible unit cost; or a product of extremely high purity may be needed to meet certain stringent user specifications. In each case the basic requirements are arbitary in the sense that they are likely to prevent the fullest possible utilization of the natural resource.

Under normal commercial conditions one of two requirements should predominate:

(i) maximum extraction of all values contained in an ore body so that the natural assets are fully converted into saleable products.

(ii) integration of all components of the production sequence so as to result in a maximum return on the investment.

The two requirements are rarely compatible and varying compromises have to be adopted in the light of specific circumstances. In the case of ores carrying high price commodities in low concentrations, emphasis on highest possible recovery of the values (95–98%) is especially important and the grade of the primary concentrate is of lesser concern. Since product tonnage is likely to be relatively small and the unit value high, sophisticated processes can be adopted for further refining of the product.

In the case of ores carrying low price commodities in high concentrations, it is usual to aim for a high grade of the primary concentrate to meet market specifications. A consequential reduction in the recovery of values (65–75%) is generally acceptable.

If a mineral resource is to be utilised as a basis for regional development, the dominant basic needs are somewhat modified:

(i) the product must satisfy the technical market requirements and user specifications at all times in order to establish market confidence in the reliability of a "new" supplier.

(ii) although the operation is likely to be initiated with a high content of imported competence, it should be based entirely on well-established and known technology. Process innovations should be avoided as completely as possible, so that technology transfer may proceed unimpeded by needs for process improvements which usually plague novel techniques.

(iii) in the light of (i) and (ii) it is preferable that a processing sequence which is conventional for a given type of ore should be given a plant design which is relatively tolerant of operator error; which can be operated initially to emphasize the reliability of product grade; and which permits an increase in throughput and recovery with growing operator competence.

(iv) the operational sequences in mineral processing, such as crushing, grinding and separation processes, waste disposal, water management, etc. are especially suitable for the induction of novice personnel to production routines and for demonstrating the importance of operational consistency; co-ordinated materials handling; and preventive maintenance. It needs to be noted that industrial discipline does not constitute a natural way of life. Transition to it from an agricultural mode of existence is a traumatic experience after a passing phase of elation occasioned by the novelty of the work.

Mineral processing can gain relatively easy acceptance in this respect, because of its involvement with natural materials and because the consequences of good and bad operational practices are readily recognisable.

Technical Necessities for the Exploitation of Mineral Deposits

Exploration and Amenability Testing

It is a historic accident that the subject of geology was on the whole, developed as an art, or pure science, rather than as an engineering discipline directed at the appraisal of mineral resources. Even today the basic outlook of many field geologists is heavily biased towards the niceties of genetic history or classification of rocks and the mineral resources potential is not always given its deserved priority. Equally, traditional mining prospectors tend to be arbitrarily selective in searching for specific commodities instead of making systematic appraisals of the resources potential of a region. In many parts of the world, the situation has been considerably improved through comprehensive aerial surveys and, lately, through ERTS satellite information. However, detailed and systematic geological ground work is still an absolute necessity. Any aspirations for regional development based on mineral resources should be founded on a comprehensive stock of exploratory information on all potential mineral resources present. This alone permits the deliberate selection of a deposit which is most apposite for the needs of the region.

For this purpose it is most desirable that the field geologists should continually submit samples of all rocks encountered in their surveys to a mineral processing laboratory for routine inspection. This should yield an appraisal of the potential uses of all mineral components, either as sources of metals or as industrial minerals. The potential uses of common minerals such as feldspar, mica or talc should not be overlooked and minor amounts of unusual components such as rare earths or vanadium should not be ignored. Separate occurrences of several commodities might well lend themselves to a combined industrial use that might be established locally. At least, they might in combination furnish a strong incentive for serious inception of a joint development scheme which they could not sustain individually.

When a potential mineral resource has been located, amenability tests should be carried out to establish whether conventional methods can be used to concentrate the valuable constituents to meet user specifications. Only when this has shown positive results should funds be committed for the full scale exploration of the size and reserves of the deposit. If full scale exploration is undertaken, it should be accompanied by further systematic amenability studies to provide confirmation of attainable grades and recoveries. This contributes basic information for economic appraisal of the project before a decision is taken to commission full scale feasibility and process design studies. The sequence thus outlined presents a step-by-step approach which entails a progressive commitment of funds and efforts to a project. At every step the project can be abandoned with least loss, or taken further on positive evidence. Conventional assaying of prospectors' samples is not sufficient for this purpose.

The location of the mineral processing laboratory facilities depends somewhat on local circumstances. Very high competence of the personnel in recognising mineral values is the most important requirement and various combinations of local or distant services may be envisaged. For example, local facilities arranged in association with geological laboratories to perform routine preparatory work may be complemented by established organisations abroad which can carry out specialist appraisal of the samples.

It is generally not productive to regard value appraisal and initial amenability testing as a training exercise for local personnel, because it is essential to speed up this exploratory phase which must precede the firm adoption of any regional development project. Even when a likely mineral resource has been located, a minimum period of three to five years may be needed for detailed exploration of reserves and feasibility assessment. Although training of local personnel can be undertaken during that period, it is unlikely that this could contribute materially towards accelerating the progress of the project. Equally, at that stage it is not known with any confidence whether a given prospect will mature into a firm project and it is better to avoid repeated redirection of novice personnel. It is generally more productive, during the preliminary phases, to concentrate local labour on support activities and on the development of various components of infrastructure as discussed below. The proving of the basic mineral resource on which the whole project stands or falls should be treated strictly as an urgent and highly skilled task from which the attention of experts should not be diverted by secondary aims.

Process Design and Feasibility Studies

When a project has been confirmed, the next stage must be concerned with the completion of process design, preliminary plant design, associated studies of utilities and other basic information on which the final decision to adopt a regional development scheme must depend. Systematic process design can only proceed if the aims of production are clearly established in terms of type, quality and quantity of products, as well as basic constraints imposed by local conditions on the processes that are to be used.

The product parameters are usually defined by a compromise between local conditions and marketable product options. Local conditions comprise the availability and cost of capital, labour, power, water, process materials and engineering services. Few ores present processing requirements that can be met by one unique combination of process and plant design. The various alternatives generally offer an opportunity to minimise a contributing factor which is scarce or costly under local conditions. Processes may be chosen to be intensive in the use of capital, labour or power in various combinations. In most cases any one of these three ingredients may be replaced to a considerable extent by increasing the contributions made by the other two. Lack of labour in a climatically hostile location provides the most dramatic example for requiring the replacement of labour by capital intensive design based on large production units with fully computerised automatic control. Such a plant could be operated by three or four skilled workers per shift, plus a complement of highly trained technicians and "software" personnel for maintenance and control support. The cost content of such a scheme per product unit is especially high on design and capital equipment, because there is an almost total dependence on the process and plant design being correct and the equipment being reliable across the full range of foreseeable variations in operational conditions.

There are many reasons why the above example should be considered as irrelevant for regional development schemes, the most obvious being that a major reason for regional development should be the local availability of population. However, it is equally inappropriate to follow lines, familiar from the Indian subcontinent, of adopting schemes of maximum substitution of labour, to minimise the costs of capital equipment and power. Although such a scheme might employ additional hundreds or even thousands of pairs of hands it cannot lead to the growth of industry-based prosperity, because it represents the industrial equivalent of subsistence farming. The multiple use of manual labour is comparatively inefficient in terms of product quality and quantity. The latter can be usually correlated with capital investment and it is axiomatic that a useful growth rate can be attained only if both capital investment and resultant gross product are large enough. The purpose of the exercise is not to transfer people from farming subsistence to industrial subsistence of a comparable low level, but to convert them into skilled manpower capable of converting resources into profits.

The above considerations indicate that process design for the exploitation of mineral resources for regional development needs a balanced mix of the following order:

(i) capital investment must be large enough to generate a useful return for the project. In the absence of a commercial definition of the rate of return on the investment, it is best to assess the amount of gross income that is large enough and of suitable duration for the establishment of a settled local community with appropriate infrastructure. This community must be built up to be capable of accepting the transferred technology needed for exploiting the mineral resource and then of sustaining technological growth for the development of secondary industries.

(ii) initial employment must be planned in the mineral project for a labour force large enough to sustain the new local community on an adequate scale. For a mine large enough to sustain major development it is necessary to think in terms of a community of at least 20,000 people. This would yield a primary labour force of about 2,000, equal to about 600 per shift for mining, milling, transport and related functions. Full provision must be made for all attendant requirements in local administration, welfare, schooling and other services. The community must be planned for long-term growth and the initial constructional phase of establishing this community must antecede the start of mineral production. It may coincide with development and constructional periods of the mine which may extend over five or six years in the case of a larger project.

(iii) the designed power requirement must be chosen in relation to (i) and (ii), but suitably modified by the local availability and cost of fuel and power. For example, this factor might result in very different process selections for identical projects situated respectively in Arabia and in the Sahel region of western Africa. Apart from the obvious cost factor, the general policy of process selection in relation to power should be based on securing maximum product reliability through reduced dependence on operator skills.

If a developing country were fortunate enough to have to make a choice of developing one of several potential mineral resources within its territory, the question of fuel or power costs would acquire further significance. Different processes and commodities have significantly different contents of energy consumption in their cost requirements. Although absolute figures for valid comparisons are difficult to derive, one might consider comparative figures quoted by P.G. Kihlstedt at a recent O.E.C.D. conference. For treating low grade ores of iron, lead or copper, the energy consumption lies somewhere between 200 and 300 Kwh per ton treated, including mining, concentration and transportation to a nearby smelter. Metal production by smelting would require around 2,000 to 4,000 Kwh per ton treated. Thus, unless the country also is in possession of abundant and cheap sources of energy, there is a clear indication that any processes of concentration of metalliferous ores could be profitably extended to the attainment of the highest possible grades of concentrates. For every ton of slag-forming gangue so rejected the energy saving in smelting would be an order of magnitude greater than the additional energy used in processing.

On the other hand, the energy content in the production of constructional and industrial minerals such as cement, limestone, or road metal is also low compared with that of metallurgical smelting. This indicates that mineral-based structural materials should be employed to substitute for metals and generally to minimise metal use in a region where fuel is expensive. This might influence the process/plant design and hence the choice of processing methods for mineral resources development towards maximising the structural content and minimizing the metal/mechanical components of processing plant.

Pilot Plant Studies

The preferred process design resulting from due consideration of all the above factors, in conjunction with the direct technical requirements of a given ore, should form the basis for full feasibility studies. This raises the question of desirability of large scale pilot plant studies. The latter always represent significantly large investments of funds and of time, but they are recommended as quite indispensable for either commercial or regional development project purposes. The evidence of small batch or continuous tests is inevitably limited by various factors of scale which cannot be circumvented. A small sample of ore, up to several tens of tons, is inevitably limited in the extent to which it can represent the true characteristics of the run-of-mine ore. Even if steps can be taken to obtain a grade close to average, the sample is liable to be coarser or finer, harder or softer, or in some other way peculiar and different from average. In the interests of test management, the sample will have to be reasonably prepared and homogenised to suit the scale of the equipment and to facilitate repeatable measurement of test variables. Such procedure provides valuable information about the amenability of the ore to the particular system of treatment, but it is unlikely to reflect adequately the variable behaviour of the ore under conditions that will prevail in production.

If the operation is to be located in an environment of well-established industrial and engineering competence, it may be argued that the full scale behaviour is best studied in the production plant. There it may be mastered during the course of inevitable adjustments and modifications in a period of up to one year of commissioning, start-up and induction. The basis of this argument is that, provided the process and plant designs are reasonably on target, the cost of piloting the operation in the production plant during the start-up is not significantly greater than that of testing in a separate pilot plant. Furthermore, the time delay associated with pilot plant work is avoided as are the possible remaining problems of scale-up.

In the case of a development project in a region and with a labour force lacking a highly developed engineering background, the above approach can only be described as a prescription for failure. There can be no doubt that large-scale, continuous and prolonged pilot plant testing is necessary for the following three reasons:

(i) to confirm on a continuous basis various important characteristics of the ore, including grindability, liberation, middlings behaviour, concentrate and tailings grades, dewatering problems, etc.;

(ii) to establish as firmly as possible the viability of the chosen flow sheet and the grade/recovery variations;

(iii) to identify the sensitive areas where processes of preparation or processes of separation are likely to require close attention;

(iv) to assess the overall matching of materials handling, dilution, throughput rates and residence times in different unit processes which might cause flow sheet difficulties or could necessitate provision of intermediate buffer stock capacities;

(v) to provide familiarisation and training for key operators and supervisory personnel.

Comprehensive pilot plant test work requires a total throughput of several thousand tons and this factor alone would constitute good reason for locating the pilot plant on the project site, rather than at a contractor's premises overseas. Additional reasons include the desirability of assessing the effects of local climate and water characteristics; the convenience of realistic local training not only for operators but also for support labour; the possibility of testing development ore and establishing the effects of mine dilution or variable cut-off grades; the study of tailings disposal and water recovery; and the general build-up of integrated expertise in materials handling and maintenance.

In fact, the pilot plant on site forms an important tool of training and technology transfer, as well as a means of final definition of design and confirmation of viability of the operation. The size and exact design of the pilot plant must depend on the specific project. In principle, there are two basic variants:

(i) a pilot plant of limited capacity, say up to a maximum of ten or twenty tons per hour, but complete, self-contained and permanent. This could be retained throughout the capital life of the mine and be operated as a training establishment. In addition, it would serve for the assessment of process improvements, or amenability testing of new ore prospects, without causing interference in the production plant. Such a small pilot plant remains part of the feasibility study and may be abandoned at reasonable costs if the final analysis of the project turns out to be unacceptable.

(ii) alternatively, it is sometimes possible to construct a pilot plant in the form of one complete production line module which later would form part of the production plant comprising several such modules of equal size in parallel. This is largely controlled by the cost optimisation of the size of production units for a given throughput and other desired advantages of flexibility or control. This type of pilot plant is temporary and becomes absorbed into the main production plant. It provides no facility for subsequent pilot testing or development work. One advantage of this approach is that the pilot plant does not constitute a separate cost item, except insofar as its individual construction must add somewhat to the overall cost of the production plant in comparison with single phase integral construction. Another advantage compared with (i) above is the absence of any problems of scale-up or design lay-out in moving from pilot plant to production line. One major disadvantage is the large cost of a production module, especially since prior design of the entire production plant is necessary. This alternative should therefore be chosen only when no doubts remain about the final adoption or rejection of a project and where no further major design modifications are anticipated. It is sometimes overlooked that this also presupposes adequate knowledge of the variability of the ore body. It is safer by far, to use a smaller pilot plant if there is need to test the susceptibility of the flow sheet to variations in ore quality, or if there is uncertainty about the competence of any major section of the process in respect of known differences in the ore from important parts of the deposit. Re-designing a full production module is usually a very costly undertaking which is often hampered by basic features of construction and lay-out.

Feasibility Studies

Even if the project is designed for purposes of regional development and is not obliged to meet the economic constraints imposed by normal commercial terms of operation, it is nevertheless necessary to carry out a strict feasibility study to ascertain the probability of attaining the basic objectives. These objectives may be defined as follows:

(i) to be able to produce the quality and quantity of product to meet production targets and user specifications. This requires proving that ore of adequate quality and quantity can be produced for the projected life of the mine; that the process and plant design of the concentrator are capable of yielding the required quality and quantity of concentrate from the ore; that all waste products can be disposed of in an acceptable manner.

(ii) the technical conditions stated in (i) must be attainable within the costs and general economic constraints specified for the project. This applies especially to capital availability which must match the optimised size of operation. For example, a large project for producing iron ore pellets on a 'green field' site may have to include its supporting township, utilities, railway link, etc. and a capital investment intensity of US$ 75 to $ 80 may have to be allowed per ton of annual product capacity.

For an openpit copper ore project, the capital investment intensity for mining and concentrating may be as low as US$ 2 per annual ton capacity of feed. With a feed grade of, say, 1% Cu, a concentrate grade of 25% Cu and 90% recovery, this equals about US$ 55 per ton of annual concentrate capacity. The addition of a copper smelter and refinery would necessitate a capital investment intensity of US$ 750 to $1,000 per ton of annual copper capacity with a very marked effect of economy of size.

(iii) the necessary support facilities and services must be confirmed as technically feasible and economically acceptable under the limits of the project.

(iv) sufficient manpower of all categories of skills must be available to ensure effective survival of the mineral resource development. For overall production labour, 0.2 man-hours per ton of crude ore need to be allowed, divided approximately into 0.07 man-hours per ton for mining and 0.13 man-hours per ton for the concentrator.

(v) the project must be shown to be compatible with a sequential growth of industrial and commercial activity to meet the basic purpose of the development. There must be no competition for labour, capital, services or utilities.

(vi) the above indications of capital and manning requirements can be used to define the necessary size of a mining project in relation to the grade of the ore and the unit price of its valuable component.

Product Evaluation

The contained value or more correctly the somewhat lower recoverable value of the ore used for assessment of the economic balance of the project should be based on one of the following:

(i) either the fob value of the commodity which is being exported under a given contract price; e.g. a copper concentrate, iron ore pellets, manganese dioxide, refined kaolin, etc., or

(ii) in the case of a commodity produced for local consumption, fabrication, or manufacture, the local delivered price that it would obtain if the commodity were to be imported from the cheaper available source.

(iii) under (i) proper allowance must be made for costs of transportation, handling, storage and including inventory and all other charges accruing to the producer. Equally, under (ii) allowance must be made for all savings attributable to the local product compared with imports, including savings on foreign exchange and on transport from the fob port to the point of consumption. Economies on foreign exchange alone can give the domestic product a cost advantage of up to 10 or 20% to justify correspondingly higher costs.

(iv) in the case of development in a distress district the project may provide new earned income in lieu of various assistance payments from national or international sources. For assessing the initial capital requirements and the product value of the project, proper allowance should be made for any such savings. The same applies to labour re-settled on the project and previously receiving public assistance in another district. It is important that the development project should not be regarded as an instant economy in assistance budgets, but should be treated as a better form of investment of the capital towards long term growth in national product. The same reason underlies the demand for an optimum design balance between capital equipment and operating labour, in preference to a very labour intensive approach.

(v) the wider extension of this principle is too complex for detailed discussion in this paper, because it becomes difficult to draw distinctions in benefits accruing from subsidiary components of a development scheme. For example, roads, railways and other services may have to be introduced into a region to assist the development of a mine. These are of immediate wider utility and can provide strong impetus to trade, agriculture, and other regional activities. Such benefits should be taken into account when considering the "true cost" to the community of the production of mineral commodities from deposits that may be marginal or sub-marginal under normal commercial circumstances. It may only be possible to assess the true benefits in retrospect, without over- or under-estimation, when the growths in regional product and per capita income have become measurable.

Discussion Summary

In the discussion which followed the presentation of this review examples were given of process failures in various locations due to inadequate pilot-scale test-work. Mention was made of the difficulty of financing mineral processing projects and suggestions were put forward for incremental additions of standard plant to minimise financing problems. There was considerable agreement that mineral resource development should not proceed at the expense of industrial development but should stimulate it.

The criteria for attracting investment capital in general and from the World Bank in particular were reviewed and case-histories quoted from UNDP-assisted projects in Burma and Korea. It was stressed that new, untried technology usually was not suitable for developing countries. The importance of political stability was mentioned. An example was given of a mineral prospect, the development of which could not be financed locally, which was taken up by an international company, but development did not proceed because the same material was already being produced by the company elsewhere. Other examples were quoted of large projects commenced with imperfect technology which suffered cost overruns requiring re-financing and, in one case, closure. Small-scale project were regarded as having a smaller relative financial risk in many instances.

There was some discussion of the effects of inflation in halting the financing of projects previously considered feasible.

Finally it was pointed out by participants that some countries are too small and under developed to provide an internal market and had to rely on export of product and import of fuel and raw materials for processing the export commodity. Participants also commented that such countries could be given special consideration by their neighbours and suppliers and not held to ransom. It was further commented that international aid donors should look first at the impact on national benefit and not entirely at conventional economics.

PAPER 3

THE RECENT DEVELOPMENT OF NON-FERROUS EXTRACTIVE METALLURGY

by M. Turpin

IN THE VERY BROAD field of extractive metallurgy two major classes may be distinguished:

The metallurgy of unreactive metals, which has a history dating back several thousand years in many cases and generally involves low-grade ores: Cu, Ni, Pb, Zn, precious metals;

The metallurgy of reactive metals, which is much more recent and in which far richer ores are available for the principal metals: Mg, Al, Ti and the ferro-alloys of Si, Cr or Mn.

Iron occupies a very special position, which partly explains its enormous preponderance from the point of view of tonnages treated, because it is both abundant and unreactive enough to be reduced under conditions close to those of the metals of the first group, i.e. by the use of gaseous mixtures obtained by the incomplete combustion of carbon. The other main reason for this preponderance lies in the exceptional characteristics of its alloys, which can be produced in large quantity and at low cost.

It is true, of course, that the metallurgy of some metals of the first class, such as molybdenum, has been developed only recently, and that there are metals of the second class which are obtained from extremely low-grade ores, the most important being uranium.

What makes the above classification significant is that from the standpoint of economics there is a fundamental difference in the extent to which the extractive metallurgy determines the cost price of the metal derived from the ore. In the first case it accounts for only a very limited part of the cost. A very striking example may be seen in table 1, which gives the breakdown of the cost price of copper into the various operations involved. It shows that metallurgical processes represent less than one third of the total cost, including 17% for conventional pyrometallurgical operations.

Table 1: Break-down of the cost of production of 1 ton of ingot copper

Mining: Openpit, 10 million tons per year of ore containing 0.6% Cu .	14%
Ore Treatment: (Conventional concentration by flotation)	57%
Production of Unrefined Copper: Modern smelting	17%
Refining (electrolytic)	12%
Total	100%

The situation is quite different for aluminium (table 2), since in this case the production of bauxite accounts for less than one tenth of the total cost, whereas electrolysis alone represents two thirds.

Table 2: Break-down of the cost of production of 1 ton of ingot alumtnium

Production of raw materials	8.5%
Manufacture of alumina	24.0%
Electrolysis of alumina	64.5%
Smelting	3.0%
Total	100.0%

Therefore, for most of the unreactive metals, the main effort of producers has been directed towards improving the productivity of mining and ore-treatment operations. It is the remarkable progress made in these fields that has made it possible to exploit increasingly poorer deposits which are often more complex as well, at more or less constant or even decreasing cost. On the other hand, the extractive metallurgy of these unreactive metals has, until recently, developed at a rather slow rate because there was little economic pressure to encourage such development. Moreover, since investments in this industry are very heavy, all circumstances combined to favour maintenance of the status quo. More rapid progress, bringing the development of new processes, has been sparked by the recent emergence of a number of powerful forces, primarily the following:

The requirements of the fight against pollution;

The staggering rise in energy costs;

The need to turn to sources which are more varied and more difficult to process; and

The legitimate desire of producing countries for on-site treatment of their raw materials.

These forces are often contradictory, but they have the very great advantage of stimulating the imagination and giving rise to many innovations. There seems to be today, if anything, an excessive desire for change; hardly a month passes without a press announcement of a new process. In view of the diversity of ores, local economic conditions and products, this is not surprising. It becomes very difficult, however, to find any guidelines for making choices which will, as a rule, have important and long-lasting consequences.

For reactive metals, few new methods have progressed beyond the laboratory stage in recent years. Their metallurgical treatment already makes use of highly advanced techniques, and the development of a new technique requires the mobilisation of the most up-to-date methods, including in particular the use of special materials whose manufacture and utilisation are still in their early stages, such as carbides, borides or nitrides. Only large enterprises with access to the most advanced techniques can hope to produce innovations in these fields. Government subsidies may also be made available in the case of metals of strategic importance, such as titanium or zirconium.

We do not propose to give a detailed dsecription of the new processes currently proposed, or even a comprehensive list of them, since we do not claim to be familiar with all of them. Among the many available references on the subject, we recommend the survey published every year by *Mining Annual Review*.[1]

In Part One we describe what in our view are the main lines of the current development. In Part Two we attempt to formulate certain criteria for making choices which may be helpful when a processing method must be selected. Even when one technique accounts for the entire output, as in the case of igneous electrolysis for aluminium, there are still important variations between the proposed technological solutions, and hence there is a need to make choices, usually giving due consideration to the effects on a large segment of the chain of operations from ore to finished product. To continue with the example of the electrolysis of aluminium, the choice of a type of alumina and a bath composition has significant effects on the plant at which this alumina is produced and on the fluorine-bearing raw materials required (use or non-use of lithium fluoride).

Part One: Major Elements in the Recent Development of Metallurgical Processes

There are today four major elements in the development of metallurgical processes:

- The increased use of oxygen;
- The trend towards complex procedures generally but wrongly called "continuous";
- The development of metallurgical treatment using chlorides;
- The development of treatment by the wet method.

Increased Use of Oxygen

A very high proportion of the concentrates which are processed in the world are subjected to reductive smelting, with or without subsequent oxidation. The conventional types of equipment used, with variants, in the steel, copper, lead and zinc industries are the blast furnace, the open-hearth furnace and the converter. In fact, in order to obtain a proper separation of the gangue from the metal and to eliminate impurities, it is essential, for obvious thermodynamic reasons, to operate under conditions of reduction and oxidation in succession. Until recently, air was very widely used as the oxidising agent and as the combustion-supporting component for burners because it is usually necessary to provide heat to melt slags, mattes and metals.

However, air contains four times as much nitrogen as oxygen, necessitating the handling of extremely large volumes of gas, which has two results:

A significant loss of heat in the fumes escaping from the furnaces; and and severe problems in purifying these gases, especially removing the dust they entrain and the sulphur dioxide contained in many of them.

As we have said, for unreactive metals the cost of smelting is only a small part of the total cost of production of the metal, and for the most important metals treated in reverberatory furnaces or blast furnaces (Cu, Pb, Ni, Zn) the amount of energy used is not inordinate. On the other hand when sulphuretted ores are used, a quantity of sulphur which has become unacceptable from the standpoint of environmental protection is released into the atmosphere, usually in the form of sulphur dioxide. To indicate the dimensions of the problem, we may point out that in order to produce 1 ton of copper from an ordinary concentrate, 1.6 tons of sulphur is emitted, the situation for nickel is even worse, with 5.5 tons of sulphur per ton of nickel.

The treatment of these gases in order to eliminate the dust and sulphur dioxide content becomes more expensive as the volumes involved increase, since very low residual content values must be attained if, as is generally the case, the standards fix the maximum total quantities which must not be exceeded.

A standard method of eliminating sulphur dioxide is to manufacture sulphuric acid, but the gas must have a content of approximately 10% in order to be used for this purpose.

A number of processes, some dating back about 20 years, have been modified to use oxygen-enriched air or pure oxygen. Their development is especially important for the metallurgical extraction of copper and nickel from sulphuretted ores.

(a) *Flash smelter of Outokumpu Oy Helsinki, Finland*[2]. *Fig. 1.*

This process has been operating for many years, initially with air. It now uses enriched air, but what makes it of interest is that even when air is used, the gas emitted contains enough sulphur dioxide to be usable for the synthesis of sulphuric acid. It is also possible to liquefy the sulphur dioxide. An interesting variant consists in reducing the sulphur dioxide to sulphur in the part where the combustion fumes rise[3]. The gaseous sulphur at the outlet of the furnace is then condensed. This possibility is of particular interest when there is no use either for liquid sulphur dioxide or for sulphuric acid. The native sulphur is sufficiently inert to be stored as is or to serve as a filler for asphalt and be used for road surfacing with properties that seem promising, especially in very hot climates. This equipment is very flexible and can be used for the direct manufacture of blister copper.

In table 3 we have indicated some examples of recently tested copper operations for an Outokumpu furnace.

Table 3: Examples of Operation of an Outokumpu Flash Smelter[3]

	Content %				
	Cu	Ni	Fe	S	SiO_2
I. *Normal Operation with 40% Enriched air*					
Composition of the concentrate	22.7	0.11	32.1	34.3	4.4
Composition of the matte	64.1	0.85	10.6	21.5	—
Composition of the slag	1.5	0.05	44.4	1.6	26.6
II. *Direct Production of Unrefined Copper without Matte* (Pilot Unit)					
Composition of the concentrate	23.6	—	3.0	8.2	—
Composition of the unrefined copper	98.9	—	0.2	0.2	—
Composition of the slag	11.6	—	5.5	1.0	—
III. *Direct Production of Unrefined Copper with Matte* (Pilot Unit)					
Composition of the concentrate	46.0	—	3.1	16.3	—
Composition of the unrefined copper	98.7	—	0.1	0.9	—
Composition of the matte	79.5	—	1.2	16.4	—
Composition of the slag	7.4	—	4.7	0.3	—

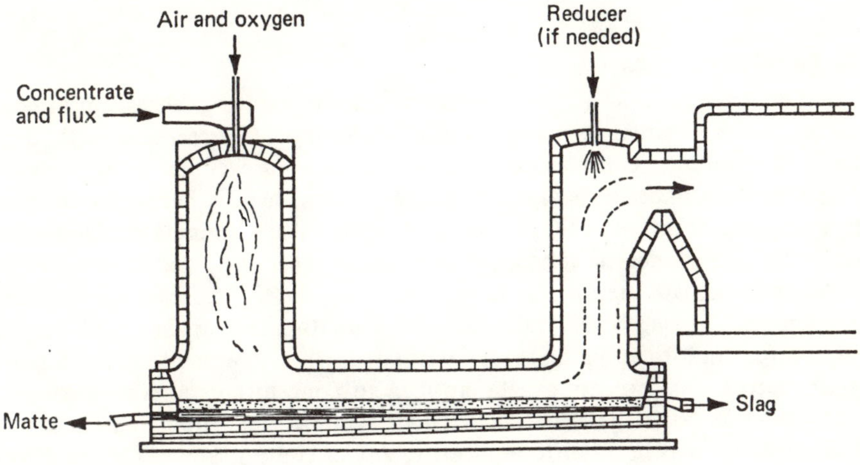

Fig. 1.

(b) *Flash Smelter of International Nickel Company, Sudbury, Canada. Fig. 2.*

This process, which has also been tested, is based on a principle similar to that used by Outokumpu but has slightly different technical characteristics[(4)]. It uses a 95% oxygen mixture to oxidise sulphide concentrates. The resulting gases contain a very high percentage of sulphur dioxide (75–80%), which is then liquefied. The mattes produced have a 50–55% copper content.

In both processes, the slag normally contains too much metal (Ni or Cu) to be discarded in that form and must be further processed: it is placed in an electric furnace, where a slightly reducing atmosphere allows the separation of a slag and a matte, after which the new slag may or may not be treated by milling and flotation.

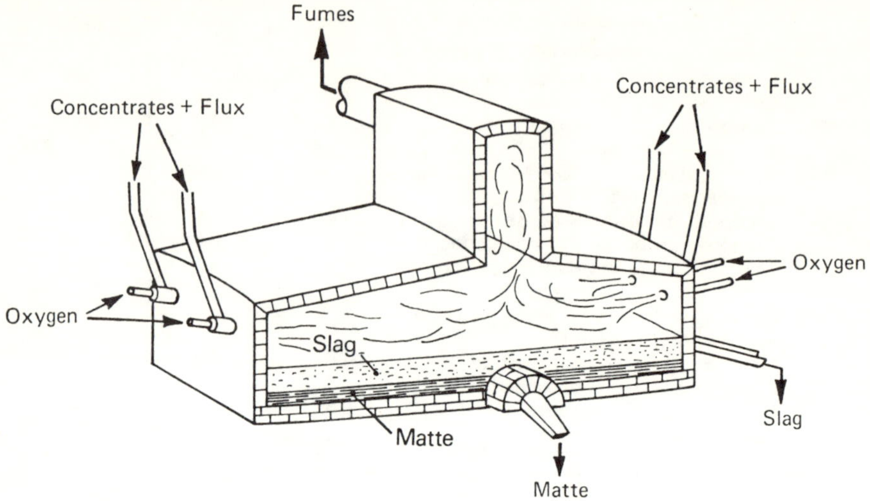

Fig. 2.

(c) *Kivcet process* (*5*). *Fig. 3.*

The Russian KIVCET process, for which Licensintorg recently granted a licence to the German company Humboldt Wedag AG uses a new method to smelt the sulphide concentrates, roasting them in a cyclone, using oxygen. As in the International Nickel Company process, the discharged gases contain a high percentage of SO_2. The second new feature consists in placing the substances in an electric resistance furnace and injecting a reducing agent in order to sublimate the tin and zinc in the slag; the tin and zinc are then recondensed. The slag is also reduced, the decantation taking place under good conditions, and it can be discarded without further treatment. According to its inventors, this process can be applied to concentrates in which copper, lead and zinc are all present. In that case, simple separation by flotation is sufficient, whereas in order to separate a copper concentrate on the one hand and a lead-zinc concentrate on the other, a carefully controlled operation is necessary. This process can also be applied to copper-tin-arsenic concentrates.

(d) *Kaldo furnace* (*TBRC: Top Blown Rotary Converter*). *International Nickel Company, Sudbury, Ontario, Canada.*

This furnace is an adaptation of the Kaldo converter, developed in Sweden about 20 years ago, for the metallurgical extraction of nickel, copper and lead. Pure oxygen is blown through the mouth of a converter, which rotates for better mixing and heat transfer, thereby ensuring better durability of the refractories, which are cooled by the bath. Nevertheless, the durability

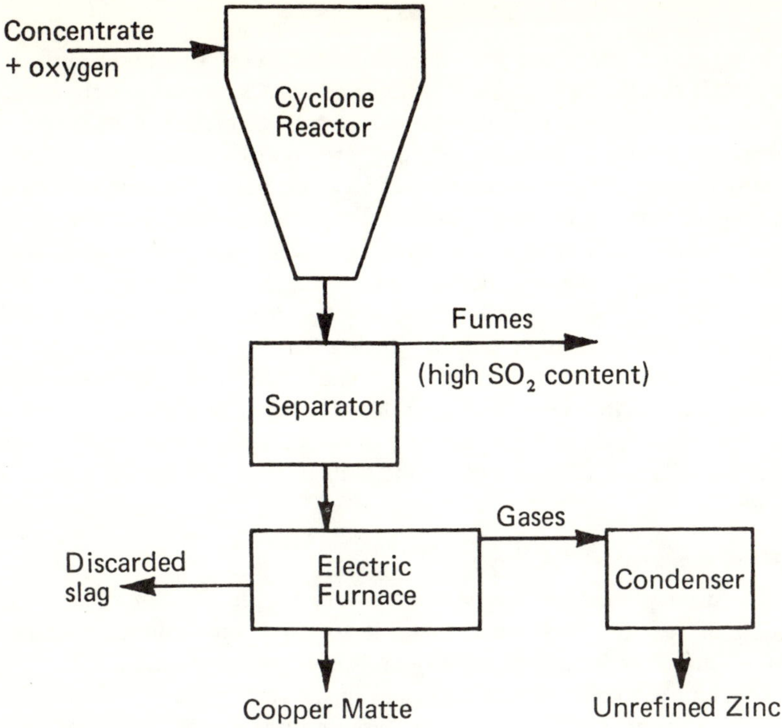

Fig. 3.

of the refractories is a critical factor in the functioning of these furnaces, since the very hot flame causes thermal shocks and significant wear.

The use of oxygen allows substantial fuel economy in the treatment of sulphide minerals because the heat given off by the oxidation of the sulphur is used in the most efficient way. It also makes possible a significantly greater output capacity per cubic metre of useful volume of the apparatus. On the other hand the process requires large quantities of oxygen, the production of which requires electrical energy, with a production cost which falls sharply as the size of the unit increases. These factors mean that the use of oxygen tends to favour large production units which can treat fairly uniform concentrates.

The use of oxygen is also widespread in continuous processes, which will be discussed next. It should likewise be noted that hydrometallurgical processes also tend to use oxygen instead of air to carry out oxidation in an aqueous medium. The kinetics of the reactions is greatly enhanced, making it possible to operate at lower temperatures and pressures.

Continuous Processes*

In order to carry out the operations necessary to obtain sufficiently pure metal, the products must be subjected to a series of widely differing physical and chemical conditions. In thermal extractive metallurgy, this involves essentially a series of reducing or oxidising operations which are normally carried out in a succession of separate units. This method has the advantage of being both simple and sure, as materials can be stored between the units, and malfunctions in any unit of the series can be corrected, at least when (as is usually the case) they are not too serious. However, the method requires a great deal of handling and involves losses both in metal and in energy. The idea of a single unit in which a number of operations could be carried out simultaneously is therefore very attractive. Several processes have been developed with a view to solving this problem. They relate mainly to the metallurgy of copper, but the principle is valid for any metallurgical process in which oxidation follows reduction; similar tests have been carried out with steel.

(a) *Mitsubishi process*[(4) (6)]. *Fig. 4.*

The principle behind the concept is very simple, as it involves a juxtaposition of reactors, making it possible to carry out reduction, smelting, decantation and reduction of the slag and subsequent oxidation. This process uses oxygen. The principal difficulty lies in the regulation of the process, as the various reactors interact. The process requires a high degree of control over the parameters and involves intricate measurement problems.

(b) *Worcra process*[(7) (8)]. *Fig. 5.*

The reactor consists of several juxtaposed zones. The slag and the metal travel in opposite directions and are run off at each end of the furnace. This process uses air, but as in the Outokumpu process, the resulting gases contain a sufficiently high percentage of sulphur dioxide to be used in the synthesis of sulphuric acid. For the time being, this process is still at the pilot-plant stage. It will probably be rather difficult to ensure proper flow of the metal and slag and good separation of the reducing and oxidising zones.

(c) *Noranda process*[(9)]. *Fig. 6.*

While the principle behind this process is similar to that of the Worcra process, the technology is simpler. Metal matter and slag are superposed according to density, and the unit is basically a converter. The charge, which is pelletised beforehand, is injected by a high-speed belt. When it

---

* The word "continuous" is confusing, since most types of metallurgical equipment, such as furnaces, operate continuously. Here, reference is being made to the continuity of separate operations carried out in the same apparatus. It generally means avoiding the effectively discontinuous functioning of the converter.

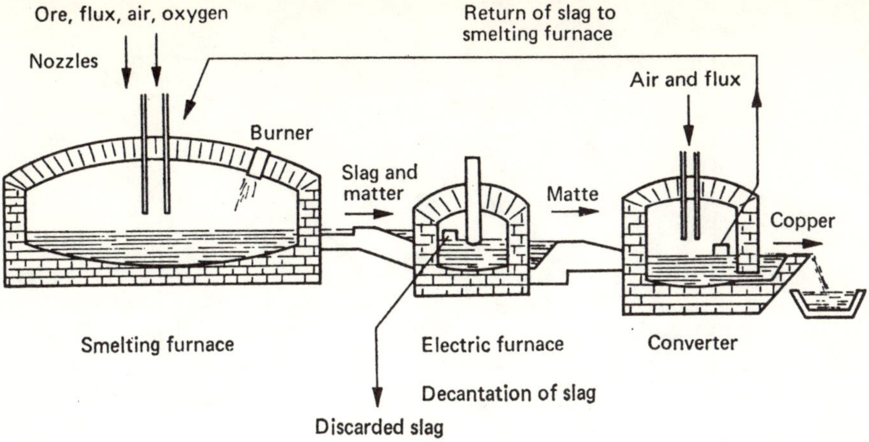

Fig. 4.

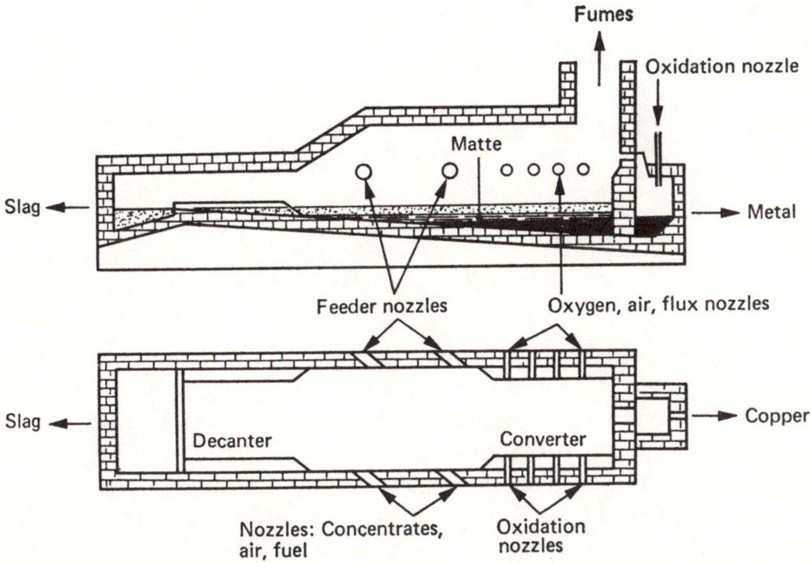

Fig. 5.

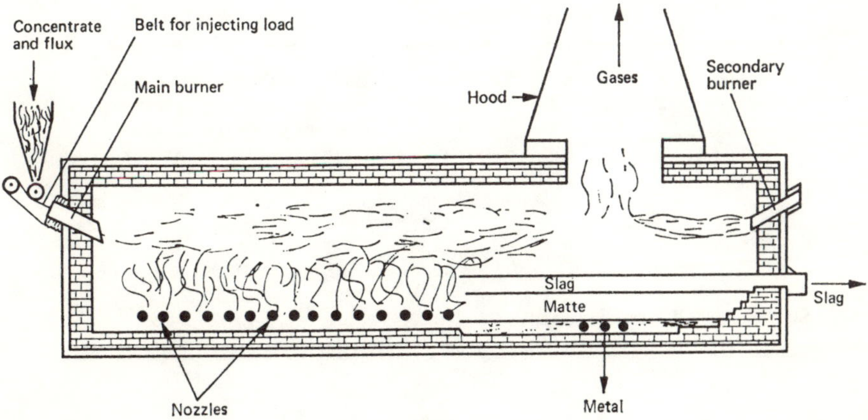

Fig. 6.

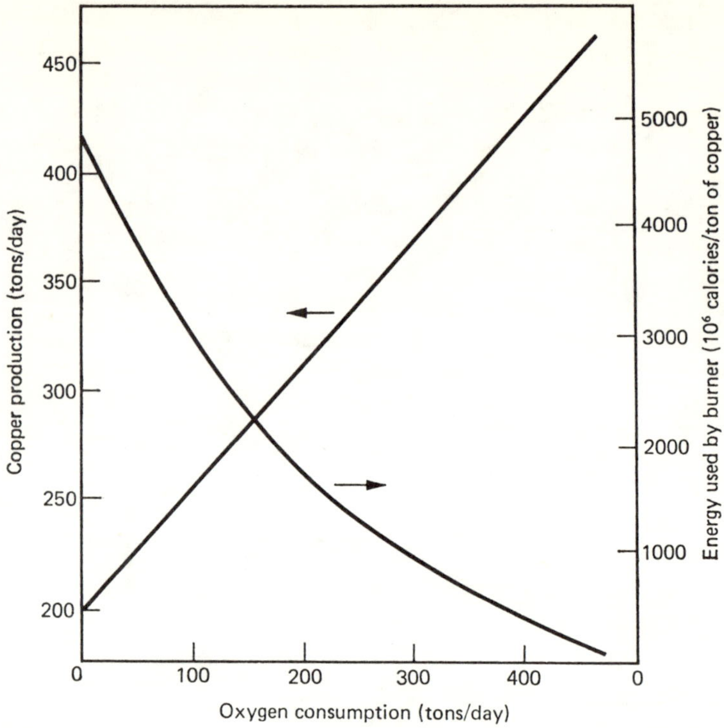

Fig. 7.

reaches the top of the slag, it is burnt partly by the oxygen from the burner but mainly by the oxygen injected into the nozzles which has not been expended in the oxidation of the matte. This matte acts as a regulator, so that there is self-regulation of the furnace. On the other hand, the slag has a very high metal content and must be treated. Given a sufficient injection of oxygen, the reactor can be autogenous, as is shown in Figure 7. The same figure shows an increase of more than 100% in the production of the unit when self-heating operation is attained, a considerable increase.

Table 4: Operation of a Noranda Converter Producing Copper[9]

At nozzles	Enriched air—35% oxygen	Daily production: 400 tons of Cu
At burners	Enriched air—50% oxygen	Energy consumption: 620×10^6 calories per ton of Cu
		Copper output of process 98.5%

Composition in % by weight

	Cu	Fe	S	SiO_2	Zn
Concentrate	24.9	29.0	32.2	4.1	4.9
Metal	97.8	0.2	2.0	—	—
Slag	12.0	34.5	1.3	21.6	—
Slag discarded after treatment	0.5	39.7	—	24.8	—

Noranda has been using this process in industrial production for two years, without oxygen, and Kennecott Copper Corporation has adopted it for its Utah division. Table 4 gives the data for operation producing unrefined copper. It is not completely self-heating. The table shows the relatively high copper content of the slag, which, at Noranda, is cooled slowly to separate all the copper and then treated by autogenous grinding and flotation. The sulphur content of the unrefined copper is higher than that reported by Outokumpu (it should be remembered that Outokumpu's is a pilot plant). The trapping of the fumes is accompanied by dilution (this avoids any escape of fumes into the working area), but their sulphur dioxide content is sufficiently high (13%) for the production of sulphuric acid.

These continuous processes are characterised by very high productivity and require the most uniform concentrates possible. The metal balance depends largely on efficient recovery from the slag and the thermal balance is highly favourable. They have all been developed to limit sulphur-dioxide pollution.

Metallurgy Using Chlorides

The use of chlorides to produce metal has been known for a long time because chlorides of metals are volatile and can be decomposed more easily than oxides.

Here we must again distinguish between unreactive and reactive metals. Chlorination is one of the possible methods for extracting unreactive metals. One fairly widespread operation consists of chlorination under heat using chlorine, hydrochloric acid or a chloride. In this case, the chloride is recovered by scrubbing of the fumes, which is followed by chemical processes in an aqueous solution. These methods are particularly well known for purifying (in the treatment of nickel oxides) or recovering low concentrations of metallic elements (in the treatment of pyrite ashes). These methods will not be dealt with here in detail, as they belong more to the field of hydrometallurgy, of which they may form a preliminary phase. We should merely point out that these operations can be carried out by means of simple and reliable equipment such as rotary furnaces.

The use of chlorides is of greatest importance in the metallurgy of reactive metals. In this case, however, it involves anhydrous chlorides which are decomposed either by igneous electrolysis in a fused salt bath (magnesium, sodium, calcium) or by metallothermic treatment using another metal (titanium, zirconium, beryllium).

When the chlorides can be obtained in a pure form without much trouble, as those of sodium and magnesium, the only difficulties are those of electrolysis. Such is not always the case, however, and the main difficulties often arise during the preparation and handling of very pure anhydrous chlorides. Many such chlorides, e.g. those of titanium, zirconium and aluminium, can be hydrolised very easily, or alternatively, in the aqueous phase, can form hydrated complexes from which the water cannot be removed. The production of metal is generally not a very selective operation, so that very pure initial substances must be used and complex purification phases sometimes become unavoidable.

While recent reports mention interesting improvements in the preparation of anhydrous chlorides, particularly in the production of magnesium, by far the most important innovations concern the metallurgy of aluminium, since this metal is by far the most commonly used of the reactive metals. At the present time, the sole method of producing this metal is by electrolysis of alumina in a cryolite bath. The alumina itself is extracted mostly from bauxite treated by the Bayer process of digestion by soda. An alternative method should be found for three reasons:

The very high energy costs, given the high energy consumption of electrolysis (about 14 kWh per ton of aluminium);

The difficult problems caused by pollution from substances containing fluorine which are given off from the tanks;

The limited reserves of bauxite, as against the practically unlimited quantities of aluminium present in shales and clays.

Two new processes for the preparation of aluminium have recently been proposed: the ALCOA process and the TOTH process.

(a) *Alcoa process*[10]. *Fig. 8.*

Alcoa recently announced that it had built a factory with a production capacity of 15,000 tons of aluminium per year using electrolysis with anhydrous trichloride in fused chloride.

The primary material in the process is pure alumina; it is currently obtained by the Bayer process, but any other source could be used. The aim is to reduce the energy consumed in the preparation of the metal, not to treat ores which are less rich than bauxite. The process consists of two phases, both of which make use of very complex technological methods. The first phase involves the preparation of trichloride from alumina by means of a reaction with chlorine and carbon. Carbon is first deposited on the particles of alumina; this is followed by chlorination, and finally, the chloride is purified by sublimation and recondensation. Some very toxic compounds, such as phosgene and certain sulphur compounds, may be formed during this phase, and it is therefore important that the loops should be perfectly sealed and that the materials should be able to stand up to the particularly severe operating conditions. The second phase is igneous electrolysis in chloride. Very detailed studies of the cell have been carried out with a view to bringing the anode and the cathode as close together as possible, thereby limiting the loss of energy by the Joule effect and maintaining the reactor at about 700°C (as against 950 to 1000°C for the tanks used today). The reactor is also sealed and uses new materials, such as silicon nitride.

The advantages of the technique developed by Alcoa are twofold:

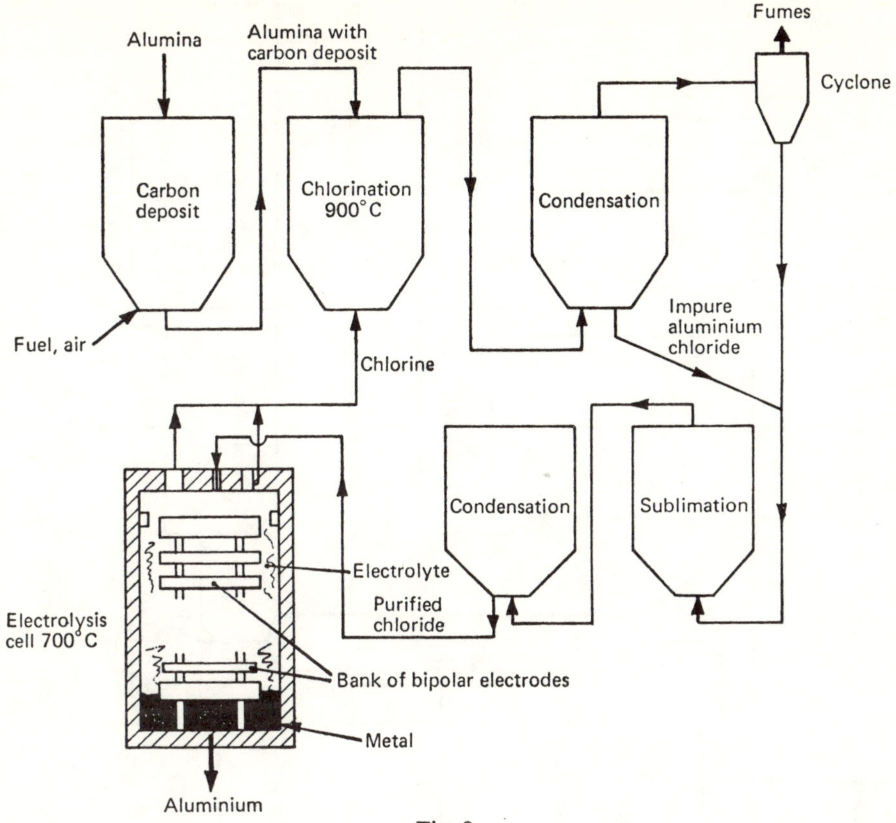

Fig. 8.

Lower energy consumption compared to the present electrolysis process, taking into account the manufacture of chloride; air-tight reactors eliminating problems of pollution.

As can be seen, the technology developed employs advanced techniques and has been carefully studied. Considerable sums have already been spent by Alcoa to develop the process, and the start of production at the plant should be followed with a great deal of interest.

(b) *The AARC TOTH process*[10]. *Fig. 9.*

This process has far more ambitious aims, since according to claims, it goes from the ore to the metal and can treat ores which contain more silica than bauxites, particularly clays. It has been given intensive publicity. It includes a chlorination phase which is similar to that of the Alcoa process but in which a mixture of chlorides is obtained, for the ores (particularly clays) are treated directly. The chloride is subsequently purified by a series of condensations and then reduced with manganese. There follows a separation of the manganese and aluminium chlorides, and then the manganese

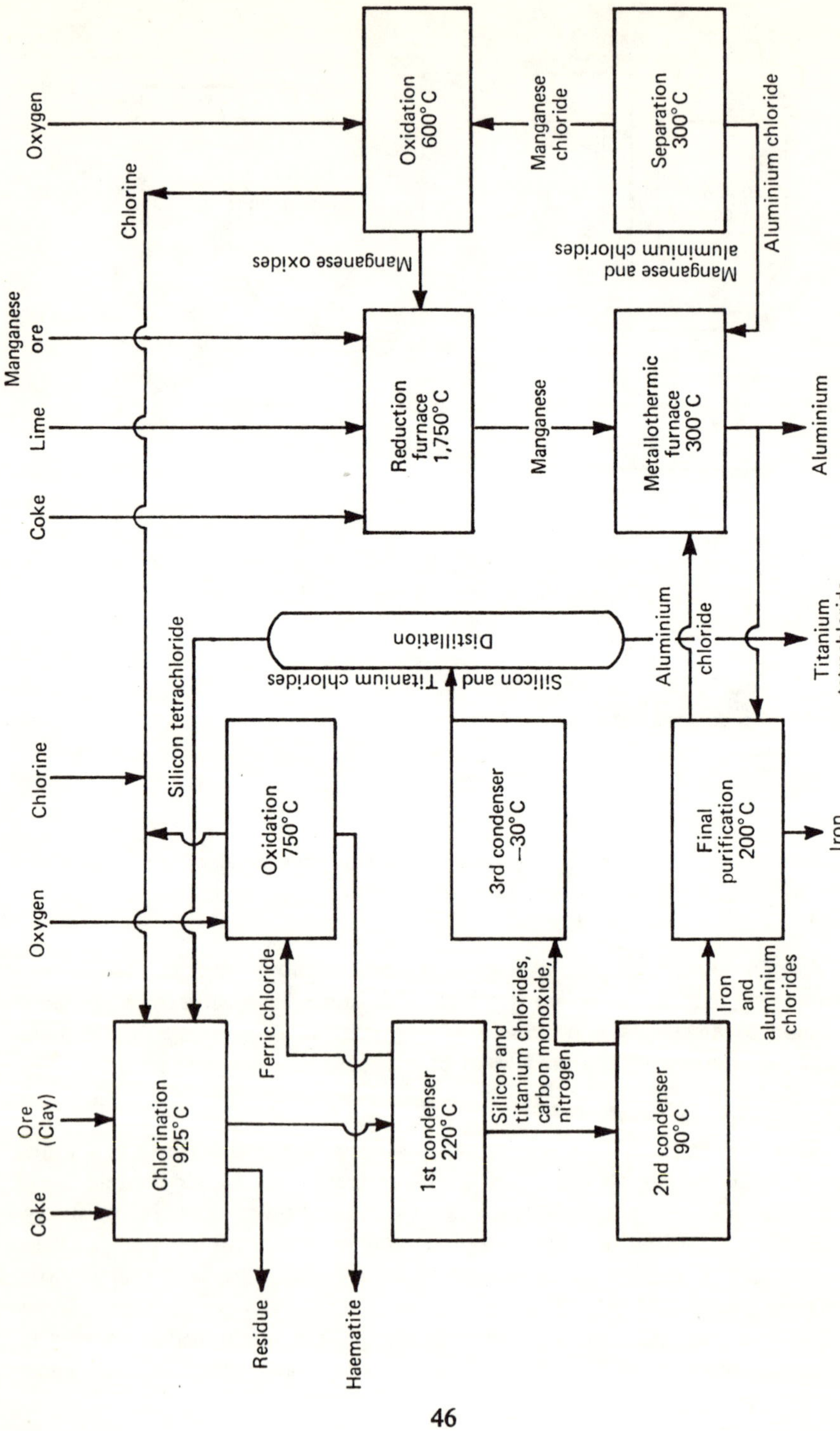
Coke
Ore (Clay)
Chlorination 925°C
Residue
Oxygen
Chlorine
Oxidation 750°C
Haematite
Ferric chloride
1st condenser 220°C
Silicon and titanium chlorides, carbon monoxide, nitrogen
2nd condenser 90°C
Iron and aluminium chlorides
3rd condenser −30°C
Final purification 200°C
Iron
Silicon tetrachloride
Silicon and Titanium chlorides
Distillation
Aluminium chloride
Titanium tetrachloride
Coke
Lime
Manganese ore
Reduction furnace 1,750°C
Manganese
Metallothermic furnace 300°C
Aluminium
Chlorine
Oxygen
Oxidation 600°C
Manganese oxides
Manganese chloride
Separation 300°C
Manganese and aluminium chlorides
Aluminium chloride

Fig. 9.

chloride is oxidised; finally, this oxide is reduced to regenerate the manganese. The literature gives few details about the techniques employed in the various phases of the process.

The presentation of the process by its promoter seems optimistic, for a number of reasons:

Chlorination is not very selective, and a complex mixture of chlorides, particularly silicon tetrachloride and ferric chloride, is obtained;

The mixture of chlorides is technically very difficult to purify and the elimination of undesirable elements also has to be provided for;

The reduction of manganese oxides with coke, as proposed, is inevitably accompanied by abundant carbide formation and does not seem to us to be sound. This would make it necessary to resort to electrolysis, so that the real advantage of the process for obtaining the very pure metal required is not at all clear.

A number of other processes have been studied in the laboratory without being followed up by industry. The technical difficulties, the cost of development and the amount of investment involved are such that any drastic change in the aluminium industry would have to be gradual.

Consequently alumina will remain a necessary intermediary in the preparation of aluminium for a long time. Because of the shortage of bauxite, various attempts have been made to treat other raw materials, particularly those very common ones in which alumina and silica are mixed and which are very poorly suited to the Bayer process, since the silica consumes soda and takes away some of the alumina in the form of a silicoaluminate. Among the new processes proposed, we may mention the Pechiney Ugine Kuhlmann H^+ process, which involves hexahydrated chloride as an intermediary and effects an acid treatment.

Development of Hydrometallurgy

A high degree of selectivity is possible with low-temperature, aqueous-phase chemical treatment; thus, pure metals can be prepared from complex solutions. This aptitude is particularly important when it is essential to produce a metal of very high purity or when certain impurities must be eliminated. The requirements of uranium metallurgy and the nuclear industry (hafnium-free zirconium and cobalt-free nickel) and the considerable resources which have been devoted to overcoming the problems encountered by the nuclear industry have made possible the development of technology which is now used for the preparation of far less noble metals. The needs of the anti-pollution campaign have aroused heightened interest in these hydrometallurgical processes, which at first sight appear very attractive.

We shall not describe the new procedures proposed, as a special paper is devoted to hydrometallurgy; we shall, however, draw attention to a number of points.

(a) *Among the advantages* of hydrometallurgy we would place first the great selectivity of the processes and the possibility of recovering all the substances placed in solution, although for successful recovery these substances must have been put in solution during the initial leaching phase. Many processes claim very high recovery rates which are not always confirmed in testing. Each process must be studied very thoroughly; it is particularly important to try out the process on the ore or concentrate to be treated, since the results may depend to a great extent on the chemical and mineralogical characteristics of the raw material.

(b) *Another advantage* is flexibility of operation. A hydrometallurgical plant can be started and stopped more readily than a pyrometallurgical plant. The investment and operating cost per ton produced is less sensitive to the size of the operation. Small units are not unduly penalised from the economic point of view. This is an important advantage when one wants to locate the metallurgical plant close to the mine.

(c) *The necessity of having at hand*, for operational purposes, certain chemical reactants—usually sulphuric acid or ammonia and sometimes hydrochloric acid—may be an advantage or a disdavantage, depending upon local conditions. In particular, a pyrometallurgical process can supply sulphuric acid, for which there is no other use at the site, and which can therefore be regarded as costing nothing.

(d) *Of greater concern* is the energy balance of hydrometallurgical processes. For example, a study of the new processes for treating copper concentrates by the wet method reveals that they are fairly heavy consumers of energy, particularly electrical energy. This is a characteristic which must be carefully examined.

(e) *Lastly*, it is not true to say that these processes are *pollution-free*. While air pollution is normally eliminated, the processes generally produce insoluble compounds, particularly sulphates, whose disposal poses problems because they often contain far more unacceptable substances, such as sulphites, thiosulphites and hyposulphates.

With these few words of warning, we nevertheless think that hydrometallurgy and electrolysis in aqueous solution will undergo significant development and that no plan for a treatment plant should be considered without taking these processes into account. The economic data, however, still tend to be uncertain, particularly as regards the size of investment.

Before going on to the second part, we should mention the processes other than those using chlorides which take advantage of the volatility of a metal or of one of its compounds. The metallurgy of lead-zinc concentrates already makes wide use of such processes. One may mention, for example the Imperial Smelting process and the distillation columns used for purifying zinc or separating cadmium. Recent developments include the processes for the recovery of tin from slags by the volatilisation of tin sulphide. One can thus treat iron-rich cassiterite concentrates in a far more elegant manner. A great deal of work has also been devoted to the development of the Mond process using nickel carbonyl. Inco has constructed a large plant based on this process but has encountered very serious difficulties and has had to adopt numerous precautions because the substance is extremely toxic. On the other hand, the final product is of very high purity.

After this brief account of the main trends in the evolution of thermal extractive metallurgy, in which it is becoming necessary to deal with increasingly difficult conditions, we shall attempt to establish some criteria that we consider important for the selection of a process.

Part Two: Some Criteria for Selection

Technological progress has made only a few processes definitively obsolete. Its main effect has been to increase the range of choices and thus to complicate the task of the decision-maker. Since the various operations are dependent on each other, the engineer is confronted with the task of constructing a fairly complex system on the basis of separate elements. We consider it advisable to avoid searching too hard for innovation unless special problems arise; it is preferable to begin by studying tested methods and choosing the best existing technologies. This advice may appear to lack originality, but it has the advantage of providing a tested solution which can serve as a point of reference and thus afford a valid element of comparison in studying variations of the basic solution or entirely new solutions. The introduction of an entirely new method of treatment is always a risky and costly solution and should therefore be resorted to only in cases where it is likely to bring major advantages.

A great many factors come into play when a choice is being made, and each individual must review the criteria he wishes to consider and must choose those which seem to him to be the most important. It is necessary to analyse the market so as to determine the type of products which are to be made: one may halt the process at the stage of a semi-finished product such as a matte or an unrefined metal, or one may carry integration to a very advanced point, including the first stage of manufacture. Further, one must take into account the availability of raw materials other than the ore which are essential for the operation of the metallurgical plant, particularly chemical products and oxygen, and it is also necessary to know what uses can be made of the various by-products which are likely to result (sulphuric acid, native sulphur, ammonium sulphate). Lastly, labour requirements, especially skilled labour, must be borne in mind, particularly when one is planning to set up a plant which is some distance away from any major urban centre and is difficult to reach, so that it might not be easy to take rapid action in the event of a breakdown.

We shall discuss three aspects of metallurgical processes, only the last of which can be easily quantified. We believe that these qualitative or at best semi-quantitative criteria are particularly important and should never be overlooked. The figures obtained from an economic calculation may lead to unsuitable choices unless a careful analysis is made of the hypotheses on which the calculations are based. Even slight modifications of these hypotheses can produce considerable variations in results and can completely change the relative merits of different processes.

1. *Reliability* is, in our view, an essential factor in choosing a new process. The process must have been proved to be capable of a high rate of operation. In a sense, this statement is self-contradictory because a completely new process does not usually have a long enough record of operation to provide full information on its reliability. This, as we have already said, means that care must be taken in introducing innovations, especially when setting up a metallurgical industry in a country which has neither long experience nor other units capable of taking over at least part of the production of the new unit if the latter encounters operational difficulties.

It is better to rely on processes which are technologically simple and which do not require very elaborate systems of control and regulation. To carry out the same operation, some processes need very special operating conditions, while others admit to much greater variation in the process conditions, for example the composition of the raw materials. Continuous processes must be studied from this angle with particular care. Complex control systems require personnel of high technical competence for the operation and servicing of the equipment, and such personnel are not always available. The ease of repair of the equipment should also be taken into account in assessing the criterion of reliability. Another important aspect of reliability is the safety of personnel. In some processes highly toxic gases are used, so that a very sensitive alarm system must be installed to avoid serious accidents. These alarms require delicate sensors which must be given preventive maintenance, and it is essential that the available personnel should be able to provide such maintenance.

Some new processes make use of very advanced technologies, either because of the materials used or because of the control systems needed. It would be unwise to try to introduce such processes in developing countries which do not have the technological infrastructure necessary for their operation.

2. *Flexibility and Selectivity*: The first requirement of a metallurgical process is that it should recover the metals contained in the ore in a satisfactory manner; this is an economic concept, since it is always possible to increase recovery if an increase in cost is accepted. In all processes, the rate of recovery depends on the nature of the concentrate or ore used, but the degree of this dependence varies from process to process. Thus, close attention must be paid to this problem, so that it will not be necessary at a later stage to make additional investments or to reject important metal values.

This problem of adaptability varies greatly with the industrial structure chosen. If the plant being envisaged is to process the ore (or the concentrate) of a single and relatively pure deposit (or group of deposits), the adaptability of the process is not very important; on the contrary, it is necessary to find a tailor-made process which is particularly adapted to the chosen ore. In such cases hydrometallurgical processes, with or without preliminary calcining or smelting, are of particular interest. Such processes make it possible to recover most of the metal values contained in the ore and do not require very large units in order to be profitable.

If, on the other hand, the products to be processed are of varied origin and have very different chemical and mineralogical properties, pyrometallurgical processes generally have an advantage, at least in producing intermediate products best defined as mattes or impure metals. One can then try to set up in one place the processing of ores of various origins, so as to use modern processes which are efficient and not too polluting.

Another aspect of the flexibility of a process is the capacity of a unit to achieve a level of production which varies in tonnage over a large range but which varies as little as possible in cost.

We would like to lay some stress on this point. Installations are designed for a nominal capacity, and the economic calculations are made on the assumption that production is to be equal to that capacity, sometimes making a correction to take into account the installation period. For many reasons, however, production may at certain times be much lower than capacity.

There is always a tendency to underestimate the duration of the installation period, especially in developing countries, where it is necessary to provide for the training of a labour force that lacks industrial experience. The training of personnel takes a long time, but it is extremely important not only for the firm but also for the country as a whole.

The metals market is subject to considerable fluctuation, and one may have to reduce production considerably without being able or willing to close the plant. Processes which make it possible to combine several units of smaller size are preferable to processes which require large production units in order to have a competitive cost price. Furthermore, the marketing of the metal usually necessitates a progressive increase in production.

3. *Energy Consumption*: The current price levels of petroleum products have altered the situation regarding the cost of energy. Table 5 shows a recent assessment[12] of the effect of the variation in energy costs on the cost of the production of different metals. The process selected has a considerable impact on this cost, even in the case of copper, where about 10 French centimes per kg can be saved by using a process of autogenous smelting which cuts down the amount of fuel used. This impact is clear in the case of aluminium and nickel produced from laterites.

Table 5: C.I.A.M. Effects of the energy crisis on the costs of production of metals[12]

Metal	Assumptions	*Total energy requirements* (kWh/ton)	*Cost before the crisis* 1973 ($/t)	*Cost after the crisis* 1980 ($/t)	*making an increase of:* cents per lb.
Al	No major hydro-electric power resources; no substitution by natural gas	34,400	173.64	307.47	6.1
Cu	Open-pit mine; metal content 0.8% coefficient of recovery: 2/1; gas furnace; no substitution by natural gas	20,200	68.15	151.68	3.8
Cu	As above except that coal and oil are partly substituted for gas	20,200	68.15	138.54	3.2
Cu	As above except for rapid smelting with natural gas and no substitution for gas	14,200	56.15	115.68	2.7
Cu	As above with substitution	14,200	56.15	108.53	2.4
Cu	As above for hydro-metallurgy; no substitution	13,850	64.70	124.75	2.7
Zn	Underground mine; metal content 7%; electrolytic smelting and refining; no substitute for gas	6,440	39.33	70.59	1.4
Zn	As above except for vertical converter	13,390	27.93	59.97	1.5
Pb-Zn	5% metal content; ISF process; no substitution	10,200	25.25	45.44	0.9 (0.6 cents per lb of Zn)
Pb	Underground; 6% Pb; conventional process; no substitution	5,441	17.96	35.40	0.8
Ni	Sulphuretted; underground; metal content: 0.5% electrical refining; only 50% of the electricity is hydro-electric	32,500	227.50	341.25	5.2
Ni	Open-pit laterite; nickel oxide by leaching; no substitution by coal or oil	120,000	535.20	952.16	18.9
Ni	Laterite; open pit; 28% ferro-nickel by electrical furnace; remainder the same	85,000	454.00	795.92	4.3
Ni	Same except coal is substituted for gas	85,000	454.00	595.12	6.4

It is also necessary to distinguish what form of energy is to be used: electricity, fuel of some kind or a particular reducing agent. This is a sphere where local conditions will play a very important role because energy which is cheap but difficult to transport may be available. It may then be more advantageous to transport the ore or the concentrate than to transport the energy needed to process it. This factor will therefore influence the choice of location for the extractive industry and may, for example, make it preferable to construct a larger unit at which raw materials from different mines would be processed together. The availability of abundant hydroelectric power is, of course, the reason for the development of aluminium manufacture in certain countries such as Norway and Canada. The existence of such resources may therefore suffice in itself to justify the installation of a metallurgical plant even if there are no local mines, especially if the ore to be processed is in plentiful supply and can be transported cheaply (by sea) to plant site.

Conclusion

Despite the recent acceleration in the rate of technological innovation, thermal extractive metallurgy is still an industry in which change is slow. Such large investments are required that the industry can embark on the development of a new technique only with the greatest caution. Often changes are made more through the further development of existing processes or through borrowing by one industry from another (typically from ferrous to non-ferrous metallurgy or vice versa) than through a genuine revolution. It is very unusual for processes to become obsolete so suddenly that it is not possible to wait for existing installations to wear out before replacing them.

The main advantage of recent changes has been to offer a wider range of choices. This broadening of the range is interesting from several points of view. It makes it possible to improve the treatment of traditional ores, taking into account the special advantages of particular local economic situations (for example, the availability of hydroelectric power or sulphuric acid, or a local market for the latter product or for native sulphur). Since extractive metallurgy is an industry with a low margin of profit, any possible utilization of by-products must be carefully studied, as must, on the other hand, any costs for disposing of the inevitable waste products of the operation.

The broadening of the range of possibilities also makes it worth while to consider processing ores which until now have been considered unexploitable because they present particular difficulties, or to treat such ores at the mining site itself by constructing a small unit specially adapted to the ore.

In choosing to develop the extractive metallurgy of non-ferrous metals to serve as the base for the economic development of a country or a region, it should be borne in mind that this industry requires large investment in comparison with the number of jobs created and that it serves a market which is often uncertain and disrupted by sudden crises.

However, on a purely technical level, it is most important to understand clearly that this industry is deceptively simple. Thermal metallurgy calls for techniques which must be reliable and durable despite the very severe conditions in which they have to operate. Competent personnel must be available for the maintenance and day-to-day repairs of the equipment, and they must be prepared to accept conditions which are sometimes fairly harsh, such as difficult surroundings and long periods of continuous work. Even if the most thoroughly tested and simplest process is chosen, it inevitably involves a high degree of technical development which affects not only the plant but also its suppliers, particularly those who supply it with its raw materials (which must be as uniform as possible), its refractories, its measuring and maintenance instruments, its ventilators, filters, pumps and the like.

The existence of difficulties which delay the start of operations, or even necessitate revision of the initial projects, should not come as an unacceptable surprise and should in no circumstances destroy the desire to continue to develop this industry.

Discussion Summary

In discussion of these issues the smelter problems of Chile, Bolivia, Burma and Turkey were reviewed in some detail and some tentative assessment was made of the various alternative new technologies and their potential for use in those countries. Most of the discussion was concerned with the newer techniques and problems of copper, aluminium and tin smelting and the possibility of recovery of iron from ferruginous bauxite residues. It was stressed that the factors to be taken into account when introducing new technology were that new processes must be completely understood by the technical personnel in the developing countries, the quality of plant design and engineering requires close attention, the quality of operators available and their training must be assured and the nature of the ore itself must be taken fully into account.

REFERENCES

1. I. C. Herbert: Extractive Metallurgy in *Mining Annual Review*, 1975 p. 253–277

2. Toivanen and Gronqvist: Nickel Refining in Finland *Canadian Mining and Metallurgical Bulletin*, 57, 653 (1964)

(3) Härkki and Juusela: New Developments in Outokumpu Flash Smelting Method TMS Paper No A74–16 AIME, New York 1974

(4) F. C. Price: Copper Technology on the Move *Chem. Engineering*, 16, 9 April 1973

(5) Kivcet process for complex ores: *World Mining*, June 1974, p. 26

(6) Suzuki, Ohyama, Shibazaki: Computer Control in Mitsubishi Continuous Copper Smelting and Converting Process. TMS Paper No A 74–14 Metallurgical Society of the AIME, New York 1974

(7) H. K. Worner, B. S. Andrews: The Worcra Continuous Copper Smelter TMS Paper No A 74–19 Metallurgical Society of the AIME, New York 1974

(8) H. K. Worner, J. O. Reynolds, B. S. Andrews: Worcra Copper Smelting, in *Copper*

Metallurgy, edited by R. P. Ehrlich, p. 108 Metallurgical Society of the AIME, New York 1970
(9) Bailey, Beck, Hallett, Washburn, Weddeck: Oxygen Smelting in the Noranda Process 104th AIME Annual Meeting, New York City, February 1975
10. J. C. Peacey and W. G. Davenport: Evaluation of alternative methods of aluminium production *Journal of Metals*, July 1974, p. 25
11. *European plant* to produce aluminium from clay by non-electric route *Metals and Materials*, 7, p. 370 (1973)
12. Interministerial Commission of Information on the Supply of Mineral Raw Materials. Ministry of Industry and Research, Paris, April 1975

PAPER 4

RECENT ADVANCES IN HYDROMETALLURGY WITH IMPLICATIONS FOR THE DEVELOPING COUNTRIES

by T. R. Scott

Recent advances in hydrometallurgy are described and where necessary correlated with the older processes which have survived the test of time. The concept of "added value" processing is examined.

Examples are given of areas where local processing can be rewarding, with special reference to metals such as aluminium, copper, nickel, titanium, zinc and uranium for large scale operations. Attention is also directed to the exploitation of smaller deposits, especially those containing less-common and high-value metals. The use of hyrometallurgy for upgrading concentrates prior to sale overseas is included. The paper ends with a discussion of the advantages and disadvantages of hydrometallurgy where comparison with pyrometallurgical processes is valid. Some areas are described where future beneficial developments in hydrometallurgical practice can be expected.

HYDROMETALLURGY IS THE recovery of metals from ores and concentrates by processes involving dissolution and selective recovery of metal values using aqueous media. Sometimes a pretreatment stage is introduced prior to dissolution (leaching) to make the starting material more reactive, while a purification step may be needed after leaching either to remove contaminants from the liquors or to upgrade intermediate products. An improvement in any one of these four stages can have dramatic effects on the economic feasibility of processing different types of mineral deposits.

Advances in hydrometallurgy can occur in the following ways:

(i) the development of processes consisting of new combinations of known metallurgical procedures,

(ii) the application of known procedures to new or different starting materials,

(iii) the use of new types of equipment or novel techniques for the treatment of ores and leach liquors,

(iv) the application of novel chemical ideas and reactions to the solving of metallurgical problems.

So-called new processes more generally fall into the first two of these categories and frequently incorporate knowledge or ideas which have existed in the open literature for a long time. New techniques and types of equipment have usually been thought of or invented long before their use became practical and have needed to wait for engineering developments or improved and cheaper materials of construction before being incorporated into "new processes". As for category (iv), it would be safe to say that 99% of the basic chemistry has been known for many years and that new chemical reactions and reagents are a comparative rarity as far as hydrometallurgy is concerned.

Because of the huge amounts of capital invested in conventional metallurgical plants, the industry is understandably conservative in its attitude to new processes and special stimuli are needed to promote their acceptance and adoption. The most powerful stimulus is the continued need for cost-minimisation in the industry, since metal prices are not rising in proportion to the increases in operating costs while the average grade of ore deposits is steadily falling. In these circumstances, new processes and techniques stand their best chance of being used if it can be demonstrated that they are cheaper than existing methods. Many such potential advances in metallurgy have remained on the shelf for years, awaiting the time when changed circumstances will favour their implementation.

The increased emphasis on environmental protection over the past decade has led to severe restrictions on the emission of sulphur dioxide by smelters treating sulphide ores, thus providing the stimulus for considering alternative hydrometallurgical processes which do not cause air pollution. Another type of stimulus has been provided by the discovery of complex ores which do not respond satisfactorily to conventional smelting operations and thus require the application of new procedures (often hydro-metallurgical) if their high potential value is to be realized.

Hydrometallurgical Processes

Some of the older operations will be mentioned as well as recent advances, thereby demonstrating how the need for the latter arose.

Pretreatment

The most common forms of pretreatment comprise flotation and magnetic or gravity upgrading, which do not fall within the scope of this paper. The other types of pretreatment are dry-way processes, such as roasting, which are necessary in preparation for succeeding hydrometallurgical processes.
Examples are as follows:

(i) Sulphide concentrates are frequently roasted to produce acid-soluble oxides by reducing the sulphide content to an acceptable minimum prior to leaching. This treatment is universally used in the electrolytic zinc industry and sometimes for mixed sulphide-oxide ores of copper.

(ii) A "sulphatising" roast under carefully-controlled conditions can be applied to sulphide ores containing copper, zinc or nickel, in such a way that iron is largely insolubilised, leaving behind soluble sulphates of the valuable metals which can be recovered simply by leaching with water. Alternatively, the procedure can be applied to oxide ores, which are roasted in an atmosphere of sulphur dioxide and air to bring about sulphation.

(iii) Heat treatment can be used with certain minerals to break down the crystal lattice and render them more reactive in subsequent leaching. The mineral beryl needs to be melted to a glass at 1,650°C to achieve this objective after which it can be readily processed to obtain beryllium salts. The lithium mineral, spodumene, needs heating to 1,000°C, to convert the naturally-occurring α-form to the more reactive β-form, before further processing. Kaolin clays, if regarded as a source of aluminium, require heat treatment at 600°–800°C, to make them readily attackable by acids or alkalis.

(iv) Roasting under reducing conditions, using an atmosphere containing hydrogen or carbon monoxide, is a feature of the Caron process, used for many years in the treatment of nickeliferous laterites. Under optimum contions, the nickel and cobalt in the ores is selectively converted to metal, leaving most of the iron combined as unreactive silicates or magnetite. The metallic material is then readily leached by aerated ammoniacal solutions. Recent improvements at the reduction stage have been achieved by adding small amounts of sulphur or salts, as in the UOP process[1] or by especially careful control of the reducing atmosphere, as recommended by the U.S. Bureau of Mines.[2] These claim to improve nickel recoveries from 70% to better than 90%, especially from the high magnesium laterites.

(v) Many thermal procedures have been proposed for chalcopyrite, including direct heating, treatment with hydrogen or steam at 800°C, heating with metallic iron or copper and reacting with sulphur at 400°C.[3] The object in all cases has been to convert the chalcopyrite to more-reactive copper sulphides for easier leaching. The writer has described another "activation" procedure which involves heating pyrrhotitic nickel sulphide concentrates at 750°C in a reducing atmosphere, thereby enabling them to be more readily dissolved in acids.[4]

(vi) "Activation grinding" in an impact or vibration mill is claimed to make chalcopyrite concentrates more reactive under acid pressure leaching conditions. This unusual approach has been successfully tested on a 3 ton per day pilot plant by Lurgi on a variety of copper concentrates.[5]

Leaching

Most leaching is done with agitation and preferably at elevated temperatures to take advantage of the considerable increase in reaction rates, which in some cases can double with each 10°C rise in temperature. Closed vessels are necessary in many instances to avoid losses of volatile reagents (e.g. ammonia) and pressure vessels are required when temperatures in excess of 100°C are used or when a high concentration of a gaseous reagent (e.g. air or oxygen) is desired.

Leaching Agents: So far as acid leaching is concerned, sulphuric acid is most often used because of its relative cheapness. Considerable attention has recently been given to hydrochloric acid, despite its greater cost, because at equivalent strengths, it is often a more active leaching agent and many procedures for metal recovery can be used with chloride solutions (q.v.). Nitric acid has received less attention, but it has the advantage of being an oxidant, a feature which is very useful in attacking sulphide minerals. With hydrochloric and nitric acids, regeneration procedures are essential, partly to avoid problems with effluent disposal, but mainly to produce recycle acid for re-use at a lower cost than fresh acid. There is so far no conclusive evidence that this latter objective can be achieved.

Alkaline leaching agents are preferred in circumstances where large quantities of acid-consuming gangue minerals are present such as iron oxides, carbonates, clays, etc. Ammoniacal solutions are favoured because of the ease with which ammonia can be regenerated. An ammonium salt also needs to be present to help form soluble complexes with the dissolved metals and to reduce the partial pressure of ammonia over the solutions. With oxide ores, ammonium carbonate is the preferred salt, since both the ammonia and the carbon dioxide can be easily recycled to regenerate the leach solutions. These solutions tend to decompose at elevated temperatures but leaching (e.g. of copper oxide ores or of metallic nickel) proceeds quite satisfactorily even at ambient temperatures in many cases. With sulphide concentrates, sulphate ion is formed as the end product of oxidative leaching, so that ammonium sulphate is the preferred salt used with ammonia. Excess of this salt formed during leaching can be sold as fertilizer or recycled as ammonia by treatment with lime. Caustic soda is less often used for leaching, since it is relatively expensive and more difficult to regenerate. It is, of course, universally used for digesting bauxite ores in the course of preparing alumina, while sodium carbonate is also used for leaching uranium ores in cases where wasteful acid-consuming minerals are present.

Of the miscellaneous reagents, chlorine water has recently been advocated as a leaching agent, especially for nickel and cobalt ores.[6] Various other oxidants, such as solutions of ferric chloride and cupric chloride, have been advocated for use in leaches where oxidation is necessary and will be mentioned below. Biochemical leaching (e.g. using bacteria to generate the leaching agent) will not be discussed in this paper, since it is mainly used in dump, heap and in-situ leaching.

Strong Acid Leaching: Most leaching operations are conducted at pulp densities around 20% by weight and at an acid strength sufficient only to dissolve the desired metal. If stronger acid is used to decompose the valuable minerals, the above procedure can often result in a high final acidity in the resulting leach liquor, leading to an expensive neutralisation step prior to the recovery stage. Consequently, certain uranium producers in Canada have done leaching at higher pulp densities (e.g. 50–75%) using 2N sulphuric acid at 70°–80°C, thereby achieving faster leaching and actually economising on acid consumption.

More recently, the ultimate step has been taken, which is to just moisten uranium ore with even stronger acid (6N) in the proportion of 1 part of liquor to 9 parts of ore.[7] This mixture is not even pasty or sticky and reaction can take place in a fixed bed at 75°–95°C in a shorter time than that required for conventional agitated leaching at lower pulp densities. The soluble uranium is eventually recovered by percolation of water through the bed. The procedure is known as "curing" and is a distinct advance in hydrometallurgical operation, resulting in major reductions in reagents and operating costs. However the technique itself is not new, since it has been the standard practice in superphosphate manufacture for many years. A similar procedure has also been advocated by Sherritt Gordon Mines Ltd. for the treatment of nickel laterites, where strong acid is needed to break down the nickel-containing minerals. In this instance,[8] the process is known as "pug" leaching, since the reaction mixture forms a pasty mass, which is later baked at 400°–500°C to insolubilise most of the iron and finally extracted with water to recover nickel and cobalt in solution.

Acid Pressure Leaching: The concept of reacting base metal sulphides with hot dilute sulphuric acid and air or oxygen under pressure was first put into practice around 1950, the objectives being to dissolve the valuable metal (e.g. zinc, copper) and convert the sulphide to elemental sulphur, which remained in the residue and could be recovered if desired. Several operations were started by the Chemical Construction Corporation in USA, working at oxygen partial pressures as high as 200 psi and at temperatures from 110°–200°C. (At the higher temperatures, sulphides were oxidised to sulphate instead of sulphur.) Technical problems arising from corrosion and the high operating pressures were frequently encountered and these small plants are now closed down. Interest was revived in the 1960's by F. A. Forward with Sherritt Gordon Mines Ltd., who demonstrated that complex zinc-lead sulphide ores could be successfully treated. Tests on a smaller scale[9] were also done by the writer on a bulk flotation concentrate from McArthur River, N.T., Australia and showed that zinc could be successfully extracted by operating at 110°C and 60 psig with oxygen instead of air, leaving lead, silver and sulphur in the residue. Pressure leaching is applicable to a variety of metallic sulphides, including those of zinc, lead, copper, iron and nickel and is especially valuable for treating mixed concentrates which cannot be economically separated into their constituent sulphides by flotation.

A related process was put into operation at Moa Bay, Cuba in 1959 and is still functioning today, emphasising that modern high pressure technology is adequate to cope with severe operating conditions.[10] In this instance, iron-rich nickeliferous laterites were leached at 200°–230°C with dilute sulphuric acid, the high temperature being required to extract the nickel and to reprecipitate any dissolved iron. An oxidising atmosphere was maintained to minimise corrosion although the oxygen played no part in the dissolution of the ore. Agitation was not done by mechanical impellers but by the introduction of high pressure (600 psi) steam. Another nickel process patented by Republic Steel Corp.[11] uses pressure digestion and oxidation on nickel laterites. The ore is first sulphidised under non-oxidising conditions, then subjected to a high-temperature, oxidising leach which forms sulphuric acid and recovers nickel and cobalt in solution. The slurry is finally autoclaved in an atmosphere of hydrogen with iron powder, thereby cementing out metallic nickel and cobalt which are simply recovered by magnetic separation.

Acid pressure digestion was also used by the writer for recovering aluminium from low-grade bauxites, clays and laterites using sulphuric acid.[12] The novel feature of this process was the recovery of a very pure basic aluminium sulphate ("hydrogen alunite") from the impure leach liquor. Subsequent calcination of this product yielded alumina and the evolved sulphur oxides were regenerated to form sulphuric acid for re-use.

A small acid pressure leaching plant in Austria has operated the Lurgi-Mitterberg process on impure chalcopyrite ores for several years, leaching at 115°C and 150–300 psi oxygen in acid-resistant autoclaves, and has been successfully adapted to recover other metals (e.g. Ni, Zn, Au, Ag) from the process streams.[5]

Leaching with Ferric and Cupric Chlorides: Ferric chloride is a powerful oxidant for refractory minerals such as chalcopyrite and numerous processes using this reagent have been proposed, especially in USA, to overcome the sulphur dioxide pollution associated with copper smelting. In the Cymet and Duval processes, finely ground chalcopyrite is digested at around 80°C to produce solutions containing iron and copper chlorides from which copper is recovered by electrolysis. The sulphide content of the ore is recovered as elemental sulphur and the ferric chloride reagent is regenerated anodically in an electrolytic cell or by treatment with oxygen or air. Excess iron dissolved from the chalcopyrite is recovered electrolytically as metal in the Cymet process or discarded as hydrated ferric oxide in the Duval process

Cupric chloride is a less powerful oxidant and can only be used on more-reactive copper minerals, such as chalcocite, bornite or the various forms of activated chalcopyrite. Since the solutions are not swamped with large quantities of ferric chloride, recovery of copper in purer form is more readily achieved. Many variants of cupric chloride leaching have been suggested and the majority use the device of reducing copper to the cuprous form prior to electrolysis, thus reducing the energy costs for electrowinning by half. During electrowinning, a portion of the copper is anodically oxidised to cupric chloride, thus regenerating the reagent for further use. A typical process has been adequately described by Cathro.[13]

Large-pilot plants (e.g. 25tpd costing US$8 million for the Cymet process) have been operated or are under construction for testing chloride leaching. Full-scale plant is or will shortly be operating at the Duval mine in Arizona. There are many problems remaining to be overcome, especially with respect to corrosion and to the build-up of impurities in the circuits.

Leaching with Chlorine Solutions: After pretreatment of nickeliferous laterites to reduce nickel to metal, aqueous chlorination has been proposed for elective dissolution of the nickel and cobalt with minimal extraction of iron.[6] Ambient temperatures can give quite satisfactory results, but quicker leaching is achieved by operating at high pulp densities, where the exothermic reaction assists in reaching a convenient operating temperature of 80°C. Typical pregnant liquors contain only twice as much impurity metals (mainly iron) as the nickel-cobalt content and are ideally suited for solvent-extraction recovery procedures. The economics of the process are said to be attractive, but large-scale testing remains to be done.

Leaching with Nitric Acid: Although nitric acid has been known as a useful leaching agent for sulphide ores for many years, it is only recently that a thorough study of a nitric acid process for chalcopyrite concentrates has been made by Kennecott Research Department.[14] Typical leaching conditions comprised 3–4 hours leaching at 90°C using 1.6 lb HNO_3 per lb of copper dissolved, operating in a continuous, co-current system. Copper extractions of 98% were obtained in locked-cycle testing, with 60% of the sulphide recovered as elemental sulphur and 80% of the gold recovered by residue cyanidation. Iron and sulphate sulphur were removed from the system as hydrogen jarosite, while acid was regenerated from NO_x gasses by contacting them with oxygen and water in stainless-steel absorption towers at pressures up to 100 psig. The major disadvantage of the process seems to be that up to 40% of the sulphide is unavoidably oxidised to sulphate instead of sulphur thereby causing increased consumption of nitric acid.

Nitric acid processes have also been suggested by U.S. Bureau of Mines for treating low-grade manganese ores and for recovery of alumina from bauxite and clay. The main problem until recently seemed to be with the regeneration of nitric acid but Kennecott appear to be quite satisfied with their regeneration technology. However, no details have been given about the economics of the process on a larger scale.

Ammoniacal Leaching: After a selective metallising reduction of oxide ores (including silicates), metals such as nickel, cobalt and copper can be readily leached at ambient temperatures with ammonia-ammonium carbonate solutions, provided that the solutions are adequately aerated to oxidise the metals. The Caron process[15] embodying these principles has long been used for treating nickeliferous laterites, while pure copper can be recovered from copper scrap by the same technique.

Another plant recovering nickel metal by leaching with ammoniacal solutions at 60°C has been operated by Inco at Sudbury, Canada since 1955.[16] The feed material is a nickeliferous pyrrhotite tailing which is first roasted to remove sulphur and then reduced to metallise nickel and cobalt prior to leaching. The leach residue is pelletised and sold as blast furnace feed.

Copper oxide ores can also be treated by ammoniacal leaching, no aeration being required in such cases. Lurgi GmbH have obtained recoveries of 85–99% from a number of different ores and claim that the process is economically feasible with ores down to 2% Cu.[17] A small commercial plant of this type is currently operating at Burra in South Australia and the procedure appears to be well-suited to relatively small deposits.[43]

Ammoniacal Pressure Leaching: Ammoniacal leaching of nickel sulphide concentrates has been successfully conducted on plant scale by Sherritt Gordon Mines Ltd.[10] since 1954 and a second plant is now operating under licence in Western Australia on local concentrates produced by Western Mining Corporation. The operation is conducted under pressure conditions, viz. 100–150 psig with air as the oxidant, and at 80°C. Nickel and cobalt are dissolved as the amine sulphates, sulphide is eventually converted to ammonium sulphate, while iron is eliminated as insoluble ferric oxide. The alkaline solutions and mild operating conditions present very few corrosion problems. The process embodies a combination of elegant chemical reactions, many of which had not previously been used on plant scale before and represents the first successful application of hydrometallurgy to the treatment of nickel sulphide ores.

More recently the Anaconda Company has promoted the Arbiter Process, which was designed in the first instance for treating copper sulphide concentrates.[18] Operating temperatures (70°–90°C) and the ammonia-ammonium sulphate leach liquors are similar to those used in the Sherritt Gordon process, but operating pressures are only 5–10 psi, using oxygen instead of air.

The novel feature of this process is to use very finely-ground material and extremely vigorous agitation, the objective being to dissolve iron temporarily in the ferrous form and to convert it to ferric oxide away from the surface of the sulphide particles, so that it does not form a protective coating and slow down further reaction. A full-scale plant is currently operating in Montana, USA, on what are essentially chalcocite concentrates, recovering copper metal from the leach liquors by solvent extraction and electrolysis. The process is also claimed to be successful in treating chalcopyrite, which is the more common but less tractable source of copper, and pilot-plant testing with this feed material has been successfully completed. Since copper smelters in America have experienced great difficulty in complying with regulations regarding sulphur dioxide emission, the advent of this new hydrometallurgical approach has been most welcome and high hopes are held for its success on full-scale plant with chalcopyrite concentrates.

Recovery of Metal Values

Recovery processes include selective precipitation of insoluble compounds of the desired metal (e.g. sulphides, hydroxides, basic sulphates, carbonates), crystallisation, gaseous reduction to metal with hydrogen under pressure, cementation on iron or other electro-negative metals, electrolysis, solvent extraction and sorption on ion-exchange resins or active carbon. The methods are so numerous and varied that emphasis here will be on relatively new processes or novel modifications of older processes.

Selective Precipitation of Compounds: By progressive neutralisation of leach liquors metallic hydroxides can be precipitated selectively in those cases where there is a large pH gap between the value applying for the desired metal and the pH at which unwanted impurities also precipitate. Examples of this technique are the selective precipitation of hydrated titanium oxide from slightly acid liquors containing large amounts of ferrous sulphate or the formation of bayerite (aluminium hydroxide) from impure alkaline solutions containing caustic soda and sodium carbonate, sometimes using CO_2 to control the pH. Many other separations which are theoretically possible cannot be used in practice because the hydroxides produced are gelatinous and difficult to filter and wash free from impurities.

Better separations are sometimes attained with basic salts. In a process[12] devised by the writer, a crystalline, readily-filterable basic sulphate of aluminium can be precipitated by hydrolysis from quite impure leach liquors at temperatures above 200°C. A readily-filterable basic sulphate of zirconium can also be recovered from acid solutions containing much silica by treatment at 100°C. Nickel is recovered as a relatively-pure basic carbonate from ammoniacal solutions by the controlled removal of ammonia by steam.

Because very great differences exist between the solubility products of metallic sulphides, many useful separations can be made from acid solutions using H_2S as the precipitant or from salkaline solutions using sodium or ammonium sulphides. By appropriate control of pH and temperature, the addition of seed material and by controlling the rate of addition of the precipitating agent, products with good filtering qualities can be obtained. Nickel and cobalt sulphides were precipitated from leach liquors in this way by the Freeport Nickel Co. in 1960, while selective removal of cobalt from the bulk of the nickel in ammoniacal liquors can be achieved because of the lower solubility of cobalt sulphide.[10] Likewise, in the Sherritt Gordon process, copper impurity can be precipitated from nickel-rich liquors because of the low solubility of the copper sulphide. In a recent process, Bussell *et al*[19] have used activated pyrrhotite as the reagent to recover nickel from highly contaminated leach liquors by operating at 100°C in closed vessels. The solid product from a dilute nickel solution (6 gpl Ni) is an enriched sulphide containing 35–40% nickel (also any cobalt) while the large amounts of impurities, especially iron and magnesium, remain in the filtrate.

Crystallisation: This technique has mostly been used for the separation of salts from brines by carefully controlled evaporation. A novel application of known chemistry is to crystallise the monohydrates of certain metallic sulphates (e.g. zinc, nickel) by heating their solutions to 180°–200°C where the solubility is very greatly decreased. In the hydrometallurgical treatment of nickel matte[20] by the Falconbridge Matte Process, crystals of pure nickel chloride are separated from concentrated solutions by the addition of HCl gas which reduces the solubility. Ferrous chloride was likewise recovered from solution in the Peace River Iron Process. Fonseca has recently proposed a copper leaching process in which pure copper ammine sulphate is crystallised from solution by adding ammonia.[21] In most hydrometallurgical procedures, however, solutions are not sufficiently concentrated to make use of crystallisation techniques.

Gaseous Reduction to Metals: In the Sherritt Gordon process, nickel is recovered from ammoniacal leach liquors by reduction with hydrogen at 400 psig and temperatures around 200°C. This operation is conducted batchwise, introducing fresh solution as nickel is depleted in the previous batch of liquor and eventually withdrawing nickel pellets after as many as 50 "densifications". The possibility exists here of separating one metal from another in solution if their reduction potentials are sufficiently far apart and in fact only 4% of the cobalt present in Sherritt liquors is coprecipitated with the nickel. Other gases (e.g. carbon monoxide) have also been studied for precipitating metals from solution.

Copper powder has been recovered by hydrogen reduction both from ammoniacal solutions produced from copper scrap or from acid sulphate solutions,[22] while Outokumpu Oy has produced cobalt by a similar technique at its plant in Finland.[23] Dasher has gone on record as saying that hydrogen reduction of copper is not competitive with electrowinning.[49] The use of carbohydrates (e.g. sugars, starch) as reducing agents has been suggested for producing copper powder from acid solutions at lower pressures and temperatures than those required for gaseous reduction.[24]

An alternative procedure is to treat slurries of nickel hydroxide or pure basic nickel carbonate with hydrogen under pressure in order to recover nickel metal powder. The ultimate process, so far only tested on laboratory scale, is to employ direct hydrogen reduction on copper-loaded organic phases derived by solvent extraction, but the major difficulty is to ensure that the organic extractants are not decomposed during the autoclave treatment.

Cementation: The oldest example of cementation is the recovery of copper from impure solutions by the addition of scrap iron. Copper, being strongly electro-positive relative to iron, is deposited as a powder and an equivalent amount of iron goes into solution. The product usually grades between 60–80% copper and requires smelting to achieve further purification.

Other electro-positive metals (e.g. lead, bismuth) can also be cemented onto iron, while a standard method for recovering silver and gold from cyanide solutions is cementation on zinc. A recent advance has been made with a process patented by the Republic Steel Corp., in which nickel is cemented out onto iron powder from highly-impure leach pulps.[11] The operation is carried out at 100°–150°C in an atmosphere of hydrogen and the metallic product containing 20–50% Ni (the grade depending on the excess of iron used) can be recovered by magnetic separation from the treated slurry. Bitzer has also described a process for cementation-in-pulp for copper metal.[25]

Cementation is a relatively simple and cheap recovery process, but technical improvements are needed to obtain higher grades of product and considerable research is still required to understand the mechanisms involved so that better control can be achieved.

Electrowinning: The process is very widely used for the recovery of the metals zinc, copper and nickel on a large scale, for iron to a lesser extent and on a small scale for silver, cadmium, etc. In most instances an insoluble anode of lead, graphite or titanium is used and the metal plated out on starter sheets at the cathode. Sulphate liquors are generally used for recovery of zinc and copper. In the latter case, up to 10 gpl iron can be tolerated as an impurity in the electrolyte, provided that some of the copper is recycled rather than stripped to exhaustion. Zinc liquors, however, have to go through an extensive purification train (q.v.) since even a few ppm of certain foreign metals can cause a decrease in the current efficiency or physical problems with zinc deposition.

Nickel can be electrowon from purified solutions, often containing mixtures of sulphate, chloride and borate anions to give optimum results. More often, however, nickel is won from soluble anodes cast from nickel matte produced by smelting. During electrolysis, nickel and metallic impurities are dissolved from the anodes which are surrounded by a diaphragm to prevent impure anolyte reaching the cathode. The anolyte is continuously pumped out of the cell and put through an elaborate sequence of purification steps before being fed to the cathode compartment of the cell where nickel is deposited. A positive head of pressure is maintained at the cathode as a further safeguard to ensure that impurities from the anolyte do not reach the cathode. Insoluble impurities, mainly sulphur, form a sludge below the anode and are periodically removed.

Electrolytic iron is too expensive to be produced in large quantity, although for some special applications it has highly desirable properties. However, the Cymet process (q.v.) was designed to produce large quantities of iron cathodes and iron powder from chalcopyrite. The chemistry is interesting, in that the electrolyte was ferrous chloride solution, part of the iron being electrowon at the cathode in a diaphragm cell, while ferric chloride was generated at the anolyte for recycling to the leaching circuit. The economics of this operation have yet to be confirmed; however it may be significant that in the related Duval process no attempt is made to recover metallic iron and ferric oxide is discarded instead.

Very recently, much work has been done on the use of particulate electrodes. Compressed lead concentrates[26] or slurried copper concentrates have been tested as soluble anodes while slurried chalcopyrite concentrates have been used as cathodes in acid electrolytes where iron is dissolved and a more reactive copper sulphide such as chalcocite is ultimately produced.[27] Developments have not yet reached plant scale and some technical problems remain to be solved, especially with the compressed anodes.

Solvent Extraction: This term has gradually changed over the years. Originally solvent extraction was mainly used in the food industry, where organic solvents were used to extract oils, etc. from solid plant materials such as crushed seeds. The same principle is still used today in hydrometallurgy to extract elemental sulphur from leach residues with kerosene or perchlorethylene, also to dissolve lead sulphate with amine reagents such as diethylene triamine, whence it can be recovered by precipitation with carbon dioxide.

From the 1940's onwards, extraction from chloride and nitrate solutions, using organic solvents immiscible with water, began to be applied, especially in the processing of nuclear fuels. Simple examples are the extraction of uranium from nitrate solutions using tributyl phosphate or the removal of ferric chloride with isopropyl ether. Unfortunately sulphate solutions, which are the most common type of liquor produced in hydrometallurgy, were not readily amenable to this kind of solvent extraction.

Eventually, however, liquid ion-exchange reagents were developed which when dissolved in inert organic solvents (e.g. kerosene) were capable of extracting metals from sulphate liquors. The first of these in the mid-1950's was the long-chain quaternary amines used for the solvent extraction of uranium; subsequently tertiary amines were developed (e.g. Alamine 336) which had better operating performance. Later, specific reagents for copper were developed (e.g. LIX 64N) which were capable of recovering copper selectively from acid liquors containing large amounts of iron and other impurities. Recent advances have now made it possible to choose an appropriate liquid ion-exchange reagent for selectively recovering many metals from impure acid or ammoniacal leach liquors (although some degree of prior purification may be necessary on occasions).

On plant scale, the great advantage of solvent extraction is that pure concentrated solutions of metal can be obtained from quite dilute, impure leach liquors. This is achieved by stripping the loaded organic reagents with relatively small volumes of stripping solution, generally dilute sulphuric acid. In the case of copper, pure solutions of 40–50 gpl Cu, suitable for the immediate production of cathode-grade metal by electrolysis, can be produced from leach liquors containing as little as 1–2 gpl Cu. Ammoniacal leach liquors are also amenable to solvent extraction using the same reagents and numerous recent processes (e.g. S.E.C. Process)[28] have been based on this sequence of operations. Space does not permit a full description of all the advances in solvent extraction in this paper but several excellent accounts of reagent development, plant operations and economics are now available.[29,31]

Another important advantage of solvent extraction is that counter-current operation with two immiscible phases is readily achieved, thus allowing high outputs from relatively small plants, with consequent economies in capital cost. The biggest development to date has been that of Nchanga Consolidated Copper Mines in Zambia which treats 12,000 gallons per minute of leach liquor in mixer-settlers of advanced design and has an output of around 200 tons per day of metal.[29] In the course of this major operation many minor problems, involving the recovery of entrained solvent from aqueous phases using coalescers, the control of acid mist evolved in the tank-house and the use of minute amounts of cobalt to improve the quality of cathodes were all satisfactorily solved.

Apart from copper and uranium, many other metals are now being commercially recovered in pure form by solvent extraction, including all the rare earths, thorium, vanadium, tungsten, hafnium, tantalum, columbium, etc., while the technique is also being used for phosphoric acid, borates, chromates, etc. In the Cymet process, it has been proposed to use a tertiary amine extractant to remove zinc from concentrated ferrous chloride solutions at low pH (approx. 1.5).[30] Ritcey has described a flowsheet for successively removing copper, zinc, cobalt and nickel from ammoniacal leach liquors using Kelex 100 as the extractant, appropriate selectivity being achieved by variations in pH, contact time and ammonium salt concentration.[31]

One minor disadvantage in solvent extraction is that feed liquors need to be completely free from all suspended material. Considerable economies in plant and operating costs could be achieved if turbid liquors or even dilute pulps could be treated directly, without incurring excessive losses of expensive solvent. Recent developments of the pulsed-column technique (instead of mixer-settlers) may allow this ambition to be fulfilled and it is understood that test installations run by the Canadian Atomic Energy Authority at Beaverlodge have operated successfully on their pulps.

Ion Exchange on Solid Resins: Ion-exchange resins, consisting of stable organic polymers containing numerous active sites for exchange of cations and anions, were first developed for water treatment. Base-exchange resins usually incorporated sulphonic or carboxylic acids while anion exchange incorporated quaternary or tertiary amines. These water-insoluble media were prepared as small beads which were packed into a column through which solutions could be passed up or down. For example, sodium and magnesium could be removed from brackish waters by passage through a column of base-exchange resin in the hydrogen form. After a back-wash, impurities were then stripped from the resin with weak acid, which simultaneously regenerated the resin to the hydrogen form for the next cycle.

Application of this principle to hydrometallurgy came in the late 1940's[32] when it was found that uranium from sulphuric acid leach liquors formed an anionic, sulphate complex which could be absorbed selectively on a strongly anionic resin and recovered by elution with an acidified solution of sodium chloride. Uranium hydroxide was then precipitated from the eluent by adjusting the pH to 6.5 with a neutralising agent such as MgO. Many variants of this general procedure have since been introduced. As an innovation, it was most timely since no solvent extraction process for sulphate solutions was then known and it provided an excellent means of concentrating uranium from the very dilute, impure leach liquors being produced from South African gold-mining tailings. For the first decade of uranium production the iron-exchange resin technique was indispensable, but thereafter was gradually displaced by solvent extraction because of the simpler engineering and higher capacities available with such systems.

More recently, events seem to be favouring the use of iron-exchange resins once again, in two main directions. Firstly, solid resins have always been potentially compatible with dilute leach pulps (thus eliminating the need for costly filtration) and technological advances in the design of continuous resin-exchange equipment have made the application of R.I.P. ("resin-in-pulp") commercially attractive. Slater has described the use of moving beds for dilute pulps (1% solids) and of fluidized or pulsed beds for thicker slurries.[33] The present technology of resin-in-pulp seems to be well ahead of that for solvent extraction in pulp.

The two main disadvantages of solid resins have been their low capacities (generally not more than 2 equivalents of metal per litre of wet, settled resin and the slow-diffusion of metals to the active sites on the resins. The latter problem has been greatly diminished by the development of more permeable resins and the further possibility now exists of using solvent-impregnated resins to increase the capacity by at least an order of magnitude. Warshawsky[34], amongst others, has recently described techniques for impregnating resins with metal-selective solvents, using the newer polymetric absorbents of the "macroreticular" type, which have very large surface areas (e.g. 100 m^2/g). Provided that solvent losses are not excessive in practice, the advantages of these resins are very great. Instead of trying to prepare resins with incorporated functional groups, the inert polymer can be used as the carrier for a wide variety of chelating reagents (specific for a particular metal) which can comprise as much as 30–40% of the bead weight. Bayer AG has patented this type of resin under the name "Levextrol" and describe the technique as "extraction chromatography for inorganic compounds". The loaded resin beads are said to behave kinetically as if they were droplets of liquid extractant of the same size.

Another development of interest is the concept of magnetic ion-exchange resins of minus 200 mesh size.[35] These contain about 10% by volume of magnetic iron oxide and are self-flocculating under quiescent conditions. For rapid absorption from solutions, the flocs are readily dispersed by agitation and then function as finely divided particles. Such resins have obvious application in the treatment of leach pulps, since they can be selectively withdrawn from the system by magnetic separation.

Gold Recovery with Activated Carbon: Although this is an old process, the technology has been greatly improved in recent years and activated carbon is tending to replace zinc in the recovery of gold from cyanide solutions. The carbon, made from coconut shell, fruit pits, some petroleum residues, etc. efficiently absorbs the gold cyanide complex from solution and loadings of 400 troy oz. Au/ton are regarded as practical. The U.S. Bureau of Mines advocates the use of expanded-bed, multiple-compartment, columns for use with filtered liquors from which recoveries better than 99% are readily obtained. Where gold is leached from slimy material not responsive to economic liquid-solid separations, carbon-in-pulp techniques may be used with pulps containing as much as 40% solids and the coarse loaded carbon recovered by screening.

The loaded carbon is stripped with 1% caustic soda—0.1% cyanide solution at 90–95°C and can then be re-activated by heating in a kiln with limited air access at 600°C. Gold metal is then won from the strip solution by electrolysis. A recent improvement uses pressure stripping at 150°–160°C which is 5–10 times faster than atmospheric pressure stripping and uses only half as much reagent.[36] The shorter time allows the carbon inventory required in the process to be considerably reduced. Savings can also be effected by returning only a scheduled percentage of the carbon to the activation stage in each cycle, since stripped carbon can be used with only slowly diminishing efficiency for 3-4 cycles before activation becomes essential. Silver can also be recovered by the same procedure, although not quite as efficiently as gold.

Purification

Ideally, a hydrometallurgical operation dissolves the desired metal (solid residue being discarded), after which the pure metal or metallic compound is selectively recovered (leaving dissolved impurities in the discarded liquor). In practice, however, one or other of the following purification steps is required.

(i) removal of interfering impurities from the leach liquor before proceeding to metal recovery. Examples will be given below for the electrolytic zinc industry.

(ii) removal of impurities during the metal recovery operation, a typical example being the continuous purification of anolyte during the electrowinning of nickel from matte anodes.

(iii) removal of unwanted impurities from an intermediate, up-graded product recovered direct from an impure leach liquor, e.g. metals recovered by cementation.

Jarosite and Goethite Processes: The electrolytic zinc industry since its inception has had trouble with zinc ferrite, which is formed unavoidably to some extent when zinc sulphide ores are roasted to produce "calcine". During leaching of the calcine, this material is not dissolved and accumulates with the residues, which thus represent a source of unrecovered zinc as well as quantities of lead and silver. Over the past 15–20 years, several residue-leaching schemes have been developed using hot strong acid, which dissolves large amounts of iron as well as the zinc.[37] The novelty in these processes has been to remove the iron from quite acid solutions in a readily filterable form while leaving the zinc in solution. The jarosite process achieves this objective by adding a salt of sodium, potassium or ammonium, which combines with most of the iron to form an insoluble basic sulphate related to the mineral jarosite. The goethite process does not rely on a foreign additive, but depends on the precipitation of a hydrated iron oxide (FeOOH) in less acid solutions (pH 2.0–3.5), under special conditions which entail keeping the ferric ion content of the impure liquor below 1 gpl throughout the purification step. Both operations take place in hot solutions (70°–100°C) and yield liquors free from excessive amounts of iron, which can then pass on to the conventional procedures for removal of other impurities (and the rest of the iron) prior to the electrolytic revovery of zinc metal.

This iron-removal technique need not be restricted to the zinc industry. The goethite-type process is used in the Cymet and Duval ferric chloride processes, to remove some or all of the iron from solution. The writer has recently advocated the same procedure for purifying nickel leach liquors heavily contaminated with iron and has studied the losses of valuable metals in the FeOOH precipitate.[38] For nickel, cobalt and zinc, losses can be kept below 1% of the total by conducting the precipitation below pH 3.0, but serious losses of copper can occur unless the pH is held at as low a value as possible.

Treatment of Zinc Silicate Ores: Zinc silicate ores have always proved difficult to process, because of the tendency for gelatinous silica to precipitate, thus causing serious filtration problems and high losses of soluble zinc. The Electrolytic Zinc Co of Australasia Ltd has recently developed an improved process[39] which produces silica in a readily filterable form and achieves recoveries of 90% with only two displacement washes of the filter cake. An essential feature of the process is to produce a leach liquor retaining up to 25 gpl silica in colloidal solution, which does not commence to gel for 10–20 hours, a period which adequately covers the leaching period and the holding time prior to the next stage. The method has been successfully tested on a 5 ton per day pilot plant and design of a full-scale plant for treating zinc silicate ore in Thailand is now proceeding.

Purification of Upgraded Materials: Cement copper, though containing 60–80% copper on the average, always contains some residual iron from the cementation process and more or less metal has been reoxidised depending on the drying process used and the conditions of storage. Likewise, nickel cemented onto iron in the Republic Steel process contains much residual iron, some cobalt and unwanted sulphate. If these materials are not purified by pyrometallurgical means, the usual hydrometallurgical treatment is to digest in ammoniacal solutions using aeration to oxidise the metals present. In this way, iron is discarded as insoluble oxide and sulphate is diluted relative to the carbonate in the solution, while cobalt can be oxidised with sodium hypochlorite and precipitated as cobaltic hydroxide with caustic soda. The pure metals can either be precipitated as basic carbonates by driving off ammonia or preferably recovered by solvent extraction with reagents such as LIX 64 or Kelex 100, stripped with dilute acid and the pure solutions electrolysed to yield pure copper or nickel.

Hydrometallurgical Applications for Mineral Deposits

The main object of this section will be to assess the opportunities for "added-value" processing in developing countries.

Although most of the known elements have been or can be treated by hydrometallurgy, it is necessary to limit this discussion to the more important metals and minerals. Pyrometallurgy (i.e. smelting) is almost exclusively used for treating ores of iron, chromium, manganese, molybdenum and vanadium for the steel-industry, while lead, tin and bismuth are most easily recovered from concentrates by relatively simple smelting processes. In the past, copper, nickel and even zinc were mostly recovered by pyrometallurgy, but the present trend is increasingly towards hydrometallurgy, particularly with copper, since copper smelters have been the chief offenders in causing air pollution with sulphur dioxide. Our first concern will therefore be with these three metals, together with aluminium, titanium and uranium. After dealing with these, briefer attention will be given to deposits of salts and non-metallic minerals and to ores containing lesser-known metals (e.g. beyllium, lithium, barium, etc). Finally, some examples of beneficiation by hydrometallurgy will be given, because this is a useful means of getting better prices for exported concentrate.

Aluminium

Around 30 million tons of alumina are produced from bauxite ores annually, of which a significant proportion comes from refineries in developing countries (e.g. Jamaica and Surinam in the West Indies, Guyana and Brazil in South America and Guinea in West Africa). Although the Bayer process for making alumina is not new, it is one of the first hydrometallurgical operations thought of for developing countries, because of the big differences in the prices of bauxite, alumina and metal, but exaggerated claims have been made to the effect that there can be tenfold increase in added value at each of these stages and it is worthwhile to examine these claims in detail.

Starting with a good-quality bauxite deposit with minimal overburden, saleable material can probably be produced at around $10 per ton. Extra costs will be incurred at this stage if the ore has to be screened to remove clays, if it occurs in pockets instead of as a uniform deposit and if after mining the environment is restored to its original condition. After transportation to a suitable port, depending on the distance, the bauxite cost may therefore have increased to $20, and shipping freights are such that the cif landed cost at (say) a European port could be around $40.

Assuming a good bauxite (50–55% Al_2O_3), two tons, costing $80, then have to be processed to give one ton only of pure alumina, selling at around $100, although such prices are rarely quoted and vary according to agreements between the various companies. The added value of $20 is not clear profit, since allowance has to be made for the capital and operating costs of the alumina plant, which are not generally available, but a generous estimate would be that a profit of $10 per ton of alumina might be made. The alumina is then transported to the reduction plant, generally sited close to cheap power and preferably near a deep-water port. It will be noted that all North American aluminium producers are so sited. From a ton of alumina, half a ton of metal is produced by an exceedingly costly process requiring high capital investment and using vast quantities of energy. The apparent added value is around $275 (half a ton of metal at $750 made from $100 worth of alumina) but in an extremely competitive industry, after allowing for freights, capital charges, operating and marketing costs, it would be very surprising to find a real profit exceeding $20 per ton of alumina from the processing of alumina to metal. After all, the tertiary producers who make cooking utensils and beer cans from the virgin metal still have to make their profit and they are not prepared to pay exorbitant prices for metal since the consumer would not be able to afford their products.

Hence the real "added-value" profit in money terms which could be obtained from processing 2 tons of bauxite to 0.5 ton metal would be $10+ $20 = $30, i.e. $15 per ton of bauxite mined. This is very different from the impression given by comparing the price of bauxite at the mine ($10) with that of virgin metal ($750 say). The main monetary benefit-gained by processing bauxite in a developing country may come from the savings in freight charges. The improvement of infrastructure and of social and cultural amenities which accompany mining and metallurgical developments may be more important than the money profits. This whole argument applies just as much to other metals such as copper, nickel, zinc, etc.

New processes for treating bauxite, clay, shales, etc. are constantly appearing in the literature. One such is the Pechiney H^+ process, which uses hot, very strong sulphuric acid to attack the ore and then treats the concentrated solutions with HCl gas to produce hydrated aluminium chloride crystals.[42] This material is re-crystallised to remove further impurities and then calcined to give a very pure alumina product. Both the sulphuric and hydrochloric acids used in the process are recovered. It is not claimed that this process can compete with good bauxite available near ports, but that its advantages will be realised in the future if bauxite is only available at lower grades and from remote locations and is subject to high taxation before export.

Copper

Although many new ammonia leaching processes have been developed for sulphide copper concentrates over the past five years, few have reached plant scale and none have been running for long enough to make certain that unexpected operating problems do not arise. The Arbiter process has made a successful start on the copper ores at Butte (Montana, U.S.A.) but has the advantages of an established mining community, tied sources of energy and an operating smelter which can deal with any incompletely extracted residues. At the time of writing, other plants in Arizona, using new processes on chalcopyrite (which is the more difficult ore to treat) are or shortly will be in operation. If developing countries are not prepared to use established smelting processes and cope with the air pollution problems which arise, the use of ammoniacal processes on new sulphide copper deposits may have to be recommended.

Another possibility would be the R.L.E. (roast-leach-electrowin) process promoted recently by Hecla Mining Company in USA.[44] The first stage is a sulphating roast which requires strictly-controlled atmospheres and temperature plus some anti-pollution measures to deal with the SO_2 and SO_3 evolved. Sulphuric acid is also produced at the electrolysis stage and may have to be discarded if no use can be found for it in the country of origin. Incidentally, the process is not new, having been studied on pilot plant scale by C.S.I.R.O. workers[41] but it is probably the least costly of any of the copper processes involving hydrometallurgy.

Other circumstances exist, however, where recent advances in solvent extraction and electrowinning can be put to good use. The examples of Nchanga Consolidated Copper Mines in Zambia (see Section 2) and the process being set up in Zaire can be followed. Similar plants in developing countries which have deposits of oxidised-copper ores would be well worth considering. A point of interest is that such operations might well be viable on smaller deposits, using the examples of Ranchers Bluebird mine or the Bagdad Copper Corp. in USA, which are on one-tenth the scale of the Zambian plant. The leaching operation for oxide ores using dilute sulphuric acid is well established and should present no problems.

An alternative procedure for oxide-copper ores is leaching with ammoniacal-ammonium carbonate solutions at moderate temperatures to recover high purity copper oxide for sale. The operations are relatively uncomplicated, as evidenced by a small plant which was set up in 1971 at Burra, South Australia.[43] Only limited accounts of previous operations were available in the literature, but a six-month pilot-plant investigation by the Australian Mineral Development Laboratories provided an adequate basis for construction and operation of the full-scale plant.

Nickel

Although the major production of nickel in the past has come from sulphide deposits in U.S.S.R. and Canada, it is usually predicted that most of the world's future reserves lie in lateritic nickel deposits. A feature of these deposits is that they are not readily amenable to the production of concentrates by mineral dressing, so that it has become essential to process them in the country of origin.

Hydrometallurgical processes are used for the recovery of nickel from nickeliferous laterites, most of which occur in developing countries. Production is either taking place or being started in Australia, Brazil, Colombia, Cuba, Dominican Republic, Guatemala, India, Indonesia, New Caledonia, the Philippines, Venezuela and Yugoslavia. In all cases, these operations employ well known processes (e.g. the Caron process already mentioned but it is very likely that recent advances including solvent extraction and electrowinning will be introduced in new plants, using integrated processes such as the S.E.C. nickel process.[28]

Also of interest is an improvement of the Caron process which has been developed by Universal Oil Products[1] and which claims recoveries of 90% nickel even from high-magnesium ores. With such high recoveries, deposits of marginal grade could be expected to be developed profitably, and reserves in older deposits augmented. Process operating costs are assessed at around 70 cents per lb of nickel (of which 47 cents are the charge for fuel oil) and capital charges would be another 35-40 cents. Both figures depend on grade and output. A plant to produce 20,000 tons of nickel a year in Latin America was costed at $125 million.

Where platinoids occur in nickeliferous laterites it may be feasible to extract them by hydrometallurgy from the residues after extraction of nickel and other base metals.

An excellent example of the development of nickel processing in the Philippines[40], is provided by the Nonoc mine of the Marinduque Mining and

Industrial Corporation. They use the Caron process with improvements introduced by Sherritt Gordon Mines Ltd., a Canadian company which has been foremost in developing ammoniacal leaching processes. Since 1969–70, Sherritt has designed, built and operated a demonstration plant using 6,000 tons of selected ore shipped to Canada, and will supply all its know-how for nickel and cobalt recovery to Marinduque and assist in the construction and operation of its full-scale plant. The agreement with Sherritt provides for a royalty payment not exceeding 20% of net annual profits for a period of years while the Philippines government will take 5% royalty based on gross annual sales. The entire cost of the project in 1971 was estimated at $232 million for the annual production of 34,200 tons of refined nickel plus 5,000 tons of mixed Co-Ni sulphide concentrate. A major operating cost will be in drying the ore which in common with many other laterite deposits contains 25% water. Loan capital was provided mainly by overseas sources and will be financed from production, the bulk of the sales contracts for an initial seven-year period being made with two Japanese companies. The plant started operating in October 1974 and will employ 2,000 people.

Titanium

The main sources of titanium and its chief compound titanium dioxide pigment are the minerals ilmenite (50–55% TiO_2, sometimes higher) and rutile (95% TiO_2 or better). Both are recovered from beachsand deposits and as a by-product of placer tin mining. The former commands only a low price $25/ton cif but is used to make high-value titanium dioxide pigment ($700 per ton) by the sulphate process. The ilmenite is digested with sulphuric acid at 160°C to yield a mixture of iron and titanium sulphates. This is extracted with water. From the solution, after reduction with scrap iron and crystallisation of about 60% of the iron as ferrous sulphate, the titanium is precipitated by hydrolysis as titanium hydroxide which is calcined to the oxide. Disposal of the ferrous sulphate is a problem (it is not always saleable) as is also the disposal of the waste liquors. The chloride process is therefore favoured, as the chlorine contained in both the ferric and the titanium chloride products can be recovered and recycled, yielding titanium dioxide and ferric oxide, the latter being usable as iron-smelting feed. The use of rutile as a feed to the chloride process is clearly more favourable. As rutile is worth nearly $300 per ton much effort has been put into converting ilmenite to "synthetic rutile". This has been done by reduction reactions followed by leaching of the iron component by sulphuric or hydrochloric acids. Effluent problems arise similar to those in the manufacture of pigment. An alternative approach is "rusting" of the reduced iron by sea-water followed by elutriation of the "rust" but this also yields an effluent which may be objectionable.[50] There is scope, given favourable geographic and economic circumstances, for both pigment and "synthetic rutile" processes to be installed at point of production of ilmenite concentrates, as has been done in Australia, Japan and Malaysia.

Zinc

Two very important advances have recently been made in zinc hydrometallurgy. The first is residue treatment, which allows considerable extra zinc to be recovered from the residues produced by the usual dilute acid leach of zinc calcines thus increasing the overall profitability of the operation. The second advance is in the treatment of zinc silicate ores, which have previously been difficult to process. An interesting example is provided below of the steps being taken to introduce the silicate process in a developing country and illustrates the complexity of the negotiations.

The Mae-Sod orebody of rich zinc silicate ore was discovered in Northern Thailand about 1956 by the Thai Department of Mineral Resources. Although Japanese interests substantially determined the extent and quality of the deposit between 1957 and 1963, they proceeded no further because of the lack of satisfactory hydrometallurgical process. After Electrolytic Zinc Co. of Australasia had patented their improved zinc silicate leaching process[39], American interests arranged for them to do metallurgical test work in Australia and to provide technical assistance and detail on design. Later, substantial pilot plant testing was done in USA by the New Jersey Zinc Co. which has a major interest in the project. All this prior investigation will hopefully prepare the ground for building in Thailand the 50,000 ton per year electrolytic zinc plant which has recently been announced.

It must be emphasised that the major cost in the hydrometallurgical production of zinc metal is electrical energy and, unless a developing country has surplus sources of cheap energy available, such a project as that described above would be inadvisable. On the other hand, if the establishment of a new hydroelectric scheme in Thailand is feasible, this represents a flow-on development, illustrating the added advantages arising from local development of a mineral resource.

The value of zinc in the more common zinc sulphide concentrates is about 60% of that of the metal eventually marketed. To obtain this added value by local smelting or hydrometallurgy economic considerations will dictate a minimum scale of operations. This is likely to exceed 100,000 tons per year. In 1974, the cost of such a size of electrolytic zinc plant in Australia was estimated to be around $130 million.

Uranium

Ores of uranium are a good example of those which repay hydrometallurgical treatment in the country of origin. With a few exceptions, high-grade concentrates cannot be prepared by ore-dressing procedures. Sales of ore to consumer countries would not be rewarding because of freight costs.

The leaching of uranium ores is a relatively straightforward operation and can be applied even to very low grade materials (e.g. 0.5 kg U_3O_8 per tonne). Recent advances have been made in strong acid leaching with sulphuric acid and especially in the recovery of pure uranium oxide by new techniques involving either solvent extraction or ion-exchange resins in pulp. For the leaching stage, high temperatures are not essential and the chief reagents (sulphuric acid and sometimes an oxidant such as sodium chlorate or manganese dioxide) are not expensive. The scale of operations need not be large (e.g. 500–1,000 ton per year of yellow oxide is quite practicable) although economies in operation are naturally achieved by aiming for outputs of 5,000–10,000 ton per year. A recent estimate for setting up operations for producing 6,000 ton per year U_3O_8 at the Ranger deposit (N.T., Australia) was well in excess of $100 million.

Deposits of Salts

Deposits of common salt, potash, borates, iodides and bromides can produce a ready cash flow for developing countries, especially if found in either surface deposits or salt lakes favourably situated for transportation. Techniques related to hydrometallurgy, such as crystallisation, purification and chemical processing are used to produce iodine, bromine, and boric acid. It is worth noting that Israel, when a newly emerging nation, lost no time in developing the chemical resources of the Dead Sea

Less Common and By-Product Metals

Metals usually recovered as by-products in hydrometallurgical operations include cadmium, thallium and indium (from zinc ores), cobalt (from copper and nickel ores), selenium and tellurium (from copper ores) and scandium (from uranium and wolframite ores). Molybdenum is recovered not only from its own ores but also as a by-product molybdenite concentrate from copper ores, which is refined by leaching with sodium cyanide or ferric chloride. Germanium and gallium are recovered from flue dusts and gas liquors in smelters and coal-burning power stations. Niobium and tantalum are recovered from tin-smelter slags. Clearly, in establishing mines and metallurgical plants, consideration should always be given to the recovery of by-products in order to increase the value of the operation and avoid wasteful use of resources.

Metals usually recovered from their own ores include beryllium, caesium and rubidium, lithium, magnesium, niobium and tantalum, thorium and other rare-earth metals, tungsten and zirconium and hafnium. All of these require complex chemistry. In many cases the weight of acids and other reagents required exceeds the weight of the mineral substance treated so that it may pay only to transport the minerals to the industrial source of chemicals and not vice-versa.

This may inhibit treatment possibilities in some localities. However, local treatment of monazite for production of thoria and rare earths has been found economic in India and Malaysia. The production of beryllium may be worth consideration but in most cases developments are likely to await the growth of industrialisation, the local availability of chemicals as by-products of other operations and the emergence of scientific and technical cadres.

Gold and silver, of course, preceded uranium as the prime examples of metals for which economic necessity dictated local recovery by hydrometallurgy.

Hydrometallurgical Beneficiation

It is often worthwhile to remove an impurity from a concentrate before selling it overseas for various reasons, including savings in freight and especially the avoidance of penalties. A simple example is the removal of iron stains from zircon concentrates with dilute acid, which improves the appearance of the zircon sand and prevents discoloration in refractories and enamels made from it. It is not always easy to remove impurities without also attacking the valuable mineral and the general rule is that there should be a substantial difference in reactivity between the two materials before hydrometallurgical beneficiation can be useful. For example, zinc sulphides in chalcopyrite concentrates either attract a penalty or are not paid for by the smelter. Theoretically, zinc sulphide is much more readily attacked by acid during pressure leaching and should be dissolved preferentially to yield an upgraded copper concentrate. In practice, several percent of copper is also dissolved and one is left with the new problem of separating the zinc and copper in the acid solution.

Some other examples of hydrometallurgical beneficiations are given below but it must be stressed that these may not always be strictly economic and that they may involve a local pollution or nuisance which was previously exported:

(i) Copper and lead sulphides can be dissolved from molybenite concentrates by acid pressure leaching with oxygen. Less than 0.5% of the molybdenum is dissolved so that the concentrate is effectively upgraded and its subsequent processing becomes less complicated. Leaching with cyanide or ferric chloride is also employed.

(ii) Pyrrhotite can be almost completely dissolved by strong hydrochloric acid from sulphide tailings containing small amounts of the nickel mineral pentlandite, which becomes protected by a thin film of sulphur released during the decomposition of the pyrrhotite. After leaching, the unattacked residual material is sufficiently high in nickel to be blended with smelter feed[45] instead of being otherwise discarded. The leach liquors containing iron are pyrohydrolysed at high temperatures to regenerate HCl and produce purified iron oxide for blast furnace feed.

(iii) Impala Platinum in South Africa uses a modification of the Sherritt Gordon ammonia leach to remove copper and nickel from their concentrates to produce up-graded platinum-group residues which are better suited for final processing. The copper and nickel are recovered as valuable by-products.

(iv) Tin concentrates containing copper or iron sulphides may attract a penalty if sold to a custom tin smelter. The sulphides can be selectively dissolved by acid pressure leaching with air or oxygen at 110°C, leaving the unreacted tin oxide behind as higher grade and less impure material.

(v) Although small amounts of magnesium salts can often be tolerated in a zinc leaching circuit, there are occasions when it is preferable to remove them selectively. The Bunker Hill Co. (USA) pre-leaches magnesite from a zinc sulphide concentrate, using warm dilute sulphuric acid for the purpose. This beneficiation[46] overcomes the build-up of magnesium in the subsequent leaching of zinc calcine and makes a costly bleed-off unnecessary.

(vi) Another possibility is that hydrometallurgical processes such as acid leaching can be used to remove sulphur from coal and enable it to be sold for power generation without causing SO_2 pollution problems. The feasibility of this operation would naturally depend on the nature of the sulphur-containing impurity, since pyrite would only be attacked under pressure-leaching conditions.

(vii) Calcium carbonate is a common impurity in heavy mineral concentrates prepared from beach sands and is often found in scheelite concentrates which are a valuable source of tungsten. In both instances, this impurity can be easily removed by treating the concentrates with dilute hydrochloric acid, leaving the valuable minerals unattacked and upgrading the concentrates.

Advantages and Disadvantages in Hydrometallurgy

In comparing hydrometallurgical with pyrometallurgical approaches it must be noted that in general the larger the smelter the more favourable the economics of processing, whereas efficient and profitable hydrometallurgical operations can be conducted on a much smaller scale, as witness most gold cyanidation plants and Ranchers Bluebird operation in Arizona, USA (approx. 6,000 ton per year copper). Paynter[47] has indicated that many small producers in southern Africa will need to consider setting up their own hydrometallurgical facilities because of the high costs of transporting concentrates to a smelter and the increasing custom charges for smelting and refining. Hydrometallurgical beneficiation is an area especially suited to small scale operation and the future possibility of using mobile plants for smaller deposits is worth consideration.

Capital costs per ton of metal are lower for hydrometallurgy in many cases. For treating chalcopyrite concentrates in Arizona, the Anaconda Co.

has claimed that plant for the Arbiter process would cost little more than half that for an equivalent copper smelter. Moreover, smelters generally carry a higher inventory of metal, partly because of the much larger circulating loads which are tolerated.

Where hydrometallurgical processes involve electrolysis or electrowinning the consumption of energy is bound to be high so that such works are more favourably sited near sources of hydro-electric power. On the other hand, smelters can produce much of their heat requirement by the oxidation of sulphur and the use of oxygen enrichment, especially in modern flash smelters where heat recovery is also more easily achieved because of the higher operating temperatures.

Many writers claim that hydrometallurgical plants are more flexible with respect to the nature of the feed and remain efficient at lower throughputs which sometimes cannot be avoided, e.g. because of "industrial situations". Temporary shut-down of a smelter on the other hand can be disastrous. The value of hydrometallurgy in handling complex ores has frequently been stressed, especially for zinc and cobalt in copper concentrates, although one must concede the versatility of the Imperial Smelting Furnace and of lead bullion refineries. Inert materials in the feed give little trouble in hydrometallurgy, whereas in pyrometallurgy they increase slag volume and affect its nature since they inevitably react at smelting temperatures.

Where precious metals (gold, silver and platinoids) occur in base metal concentrates it is generally easier to recover them as by-products from the metal produced by smelting. However, if their total value is less than the cost of installing and operating the elaborate anti-pollution measures required to-day, it may be preferable to consider hydrometallurgical processes for recovery of the main metal and attempt to recover the precious metals from the residues by cyanidation. The reagent cost of this may be prohibitive and thus lead to abandonment of the procedure. This illustrates the economics aspect of the conflict between conservation of the environment and the conservation of resources when the avoidance of waste is involved.

Hydrometallurgical processes have been criticised for requiring more precise control, but the writer believes this can be made into an advantage by the introduction of rapid on-stream analysis and the increased use of automation. Such devices represent one of the more attractive features of the Arbiter process, for example. Pyrometallurgy on the other hand can tolerate variations in feed rate and operating temperature but, although there are advantages in automatic control, its implementation is more troublesome.

Corrosion problems can present special difficulties in hydrometallurgical plants, all arising from the fact that water is the operating medium. Even traces of chlorides in stainless-steel holding tanks or crystallisers can cause annoying failures. The progress of acid pressure leaching was hindered for a long time by corrosion problems caused by the use of cheaper but less satisfactory materials of construction. However, modern technology has come up with special materials (e.g. titanium—0.3 palladium alloy for hot ferric chloride liquors) which can cope with most situations and pay for their initial high cost by trouble-free operation. There is also promise of cheaper materials (e.g. fibreglass, resin-bonded concrete) in the near future. The majority of hydrometallurgical processes have not suffered from serious corrosion problems, as witness the proliferation of ammoniacal leaching operations (using mild steel), of electrolytic zinc plants and of uranium leaching (which can even be done in rubber-lined, wooden pachucas). Abrasion, however, is another problem which should not be forgotten, especially if solids in the pulp are so violently agitated that the protective oxide layer on the metal surface of a vessel may suffer damage. Pyrometallurgists are not without their own related problems as illustrated by the frequency of magnetite accretions in smelters, poor fluidity of slags in the treatment of titanium ores or high-magnesium nickel ores, and the corrosion of refractory liners in converters and reverbatory furnaces.

The greatest advantage of hydrometallurgy lies in its selectivity and hence also its flexibility in treating complex ores and materials from different sources. Ammoniacal leaching agents, for example, do not dissolve silica, silicates, iron oxides or carbonates which are the commoner gangue minerals in concentrates, and thus produce relatively pure leach liquors in one step. Acid reagents are somewhat less selective, but still leave many common impurities such as silica and pyrite in the residues. In fact acid leach liquors from pressure leaching of copper ores often contain so little iron that they can proceed direct to electrowinning. An abundance of selective separation procedures can now be applied to recover copper, nickel, cobalt, zinc and many other metals directly from leach liquors and are particularly useful when complex ores or bulk flotation concentrates have been used as the feed materials to the leaching circuit. Where closed circuits are used in continuous co-current leaching, only small bleed-offs (which can be treated on a much smaller scale) are needed to remove minor impurities, and the more expensive leaching agents such as ferric chloride can then be recycled indefinitely.

With respect to overall economics, opinions and conclusions will be strongly influenced by local conditions. The current literature on preferred methods for treating chalcopyrite ores[47, 48, 49] illustrates the difficulty of arriving at objective estimates.

Problem Areas

Liquid-Solid Separations: These can be troublesome in hydrometallurgical operations. Especially with fine, slimy or gelatinous materials, the capital costs for cycloning, thickening and filtration can be greater than for any other unit operation on a plant. Residues produced by leaching or subsequent procedures each require separate study of settling and filtering properties, since these can change, even in the same plant if the source material is changed or if even minor alterations are made to variables such as the temperature, the rate of addition of reagents, the degree of agitation or the kind of filter aid used. Emphasis should therefore be placed wherever possible on recovery processes which can operate in leach pulps or at least in turbid thickener overflows. Some examples of such operations have been given earlier, e.g. cementation-in-pulp, and resin-in-pulp. Some liquid-solid separations could also be made easier if closer attention were given to the grinding and especially the classification sections preceding leaching or even flotation, since it is not always necessary to have exceedingly fine particles for leaching or for the liberation of the valuable mineral. Chemical precipitates produced during recovery and purification can also be improved by special procedures.

Reagent Regeneration: Of the acids used in leaching, sulphuric acid is the cheapest and regeneration *per se* would be more expensive than fresh acid. Even so, Amax have considered regeneration in their variant of the Moa Bay process for treating nickel laterites, the object being to crystallize out magnesium sulphate produced in large quantities during processing and regenerate magnesia by thermal decomposition for re-use in the process, while simultaneously converting evolved SO_2 and SO_3 back to acid. Operations such as electrolysis, however, regenerate acid incidentally when metal ions are removed from solution and such "spent" acid can be usefully recycled in copper and zinc plants.

The advantages of hydrochloric acid for leaching need to be very considerable to justify the use of such a relatively expensive reagent. Much publicity has been given to schemes for regenerating chloride solutions by pyrohydrolysis on the grounds that regenerated acid would be appreciably cheaper than fresh acid. At the 4th Annual Meeting of Hydrometallurgists, held in Toronto in August 1974, an entire technical session was devoted to this subject and the concensus of opinion did not favour the above concept. In fact, one paper[51] concluded by saying that "at present the total cost of regeneration exceeds the purchase price of (new) 32% acid on an equivalent HCl basis". Certainly, if credits were allowed for oxides produced during pyrohydrolysis (e.g. Fe_2O_3 from iron chloride pickle liquors), regeneration schemes might have a good chance of breaking even. On the other hand, if spent liquors from hydrochloric acid leaching constitute an effluent disposal problem, then a regeneration scheme is probably the best way of dealing with the effluent despite the expense involved. Ferric and cupric chloride systems generally operate on closed cycle and regeneration of the reagent is integrated into the circuit, either electrolytically by anodic oxidation (Cymet) or by a special foam aeration column using 45 psi air.[47] The only discard is likely to be a relatively small bleed-off to prevent impurity build-up and the volume of liquor concerned would be too small to support the capital cost of a pyrohydrolysis plant. Regeneration of nitric acid has already been mentioned.

Ammonia-ammonium carbonate liquors as used in the Caron process decompose on boiling and can easily be regenerated by absorption of ammonia and CO_2 in an aqueous counter-current system. Losses of ammonia need not exceed 1% and make-up carbon dioxide can usually be provided from flue gases produced at some part of the plant. Excess ammonia can be stripped from ammonia-ammonium sulphate liquors but the crystallised ammonium sulphate can represent a problem if ready markets are not available or if it is not pure enough for use as a fertiliser. The main alternative is to regenerate ammonia by treating with lime but the economics of this procedure would vary at different locations, depending on the relative costs of new ammonia and of suitably pure lime.

Effluent Disposal: Although hydrometallurgy does not suffer as much as smelting from air pollution problems, it is much more heavily involved with the disposal of liquid wastes, arising from barren leach liquors, and of solid wastes, consisting mostly of leach residues. The enormous tailing dams of "red mud" (bauxite residues) resulting from the hydrometallurgical processing of bauxite provide an extreme example of the magnitude of the disposal problem which should be borne in mind right from the inception of a hydrometallurgical operation. Adequate planning for treatment of wastes should have a priority no lower than that for the leaching circuit or the liquid-solid separations that have to be done. Indeed, some pyrometallurgists insist that the effluent problems of hydrometallurgy represent a bigger headache than their own.

Future Developments

The topics which have been discussed above are themselves very important areas needing future development. Some others are as follows:

(i) *Process Control*: All the features associated with improved process control, such as on-stream analysis, increased automation and sophisticated control systems with fast feed-back are urgently needed for hydrometallurgical operations. A good example is provided by Anaconda's Arbiter process, where small samples of the gases in the leaching vessels are continuously bled-off, fed to an automatic gas analyser and returned to the system. According to the analyses, automatic adjustments are made to the partial pressures of ammonia and oxygen to ensure that optimum operating conditions are maintained. Pre-treatments, such as sulphating or metallising roasts prior to leaching, have been less successful in the past than they should have been, mainly due to lack of sufficiently precise and continuous control.

(ii) *Continuous Operation*: It is hoped that the future will see the elimination of batch processing wherever possible in pressure hydrometallurgy, e.g. in the Sherritt Gordon process for ammoniacal leaching of nickel sulphides, nickel metal is still produced by an awkward and time-consuming series of batch "densifications". Continuous flow allows for higher throughput, lower cost and greater benefit from automatic controls. The problem of course lies in the cost of separating solutions from solids, unless some means can be devised for leaching with solids passing downwards in a tower-type digestor with counter-current upward flow of liquors. It is known that attempts were made to apply this technique to pressure leaching of uranium ores in the late 1950's but nothing further has been heard of the results.

(iii) *Specific Extractants*: Despite the great progress made in solvent extraction in recent years, many of the separations available to-day are governed by the need to use specialised leaching agents (e.g. ammonia) or ingenious devices (such as selective loading of copper in the presence of other metals by "crowding"). There is still a great need for selective reagents which can function in impure solutions for metals other than copper. Developments of the pulsed column will hopefully eliminate the need for extreme clarification of leach liquors, possibly along lines where fine particles are made hydrophilic so that organic reagents are displaced from the surface.

Ion-exchange resins are probably entering a major new phase of development where they will complement solvent extraction. Even now they are the best proposition for recovering metals from very dilute solutions and for use in turbid liquors or leach pulps. The commercial applications of magnetic resins and of the new macroreticular resins loaded with up to 30–40% of their weight with metal-specific extractants are eagerly awaited.

(iv) *Reaction kinetics*: One criticism of hydrometallurgy commonly made is that reaction rates, being kinetically-controlled, are often slower than the equilibrium-controlled reactions occurring during smelting. The writer is not convinced that this criticism is warranted, since the residence time in a smelter is generally of the same order of magnitude as in a leaching operation. However, it is hoped that closer attention will be given in the future to identifying and perhaps modifying the rate-controlling factors in leaching operations. Probably greater use will be made of higher leaching temperatures (e.g. pressure leaching) in an effort to speed up diffusion-controlled reactions and of improved agitation techniques to ensure rapid adequate transport of reactants (especially gases) to reaction sites on solid particles.

(v) *Equipment*: If pressure leaching comes into greater use, present-day equipment is adequate for providing the necessary corrosion resistance, strength of reactors, methods of agitation and appropriate valving for input and discharge of slurries. However, it is the capital cost of such equipment which is the deterrant, so that less expensive materials of construction are urgently needed or new techniques of cladding mild steel with coherent layers of more resistant materials might be developed. Possibly new ceramics and other non-metallic materials will help in this direction.

Engineering in the solvent-extraction field is still developing and improved methods for phase disengagement and solvent recovery will surely be forthcoming. Electrowinning techniques have still plenty of scope for improvement, as witnessed by the spate of suggestions made at a recent Symposium[52] "Cell Design in Electrowinning and Electrorefining". It is also hoped that electrolytic diaphragm cells can be developed to a higher stage of efficiency.

(vi) *Complex ores*: In spite of advances in mineral processing it is almost always impossible to produce individually pure concentrates from multimineral ores. Each such concentrate will be contaminated with quantities, sometimes substantial, of other constituents of the ore. Smelters impose penalties for these so-called impurities which are often valuable in their own right. On the other hand, concentration of the values in bulk can be cheaper and more efficient. One of the most important contributions that hydrometallurgy can make in the future is to be able to process, economically, mixed oxide-sulphide ores or bulk concentrates containing several useful metals.

Conclusion

It can be concluded from the foregoing that, apart from cases where feasibility studies clearly show the advantages of hydrometallurgical over pyrometallurgical methods for major metals such as nickel and copper, there are situations where hydrometallurgy can make a unique economic contribution.

Examples of these are the cases where concentration of values for export is impossible and smelting is impracticable (gold, uranium, nickel in laterites). Other examples are deposits (e.g. of copper ores or waste dumps) which are too small, too lean, or too remote to be otherwise considered.

Also important are the possibilities raised for added value processing of major minerals such as bauxite and of relatively small-scale production of high-value materials such as rare-earths, synthetic rutile and perhaps others, as mentioned above.

Finally, although beyond the scope of this paper, the potential for recovery of metals from sea-water should be mentioned. Magnesium and bromine are already commercially produced but there is a big field for progress in this direction.

Discussion Summary

In the course of the discussion of the above, interest was shown by the participants in the problems of uranium leaching, the treatment of ferruginous bauxites, the possibilities of extraction of alumina from clays and the recovery of magnesium from sea-water.

Reference was made to a process proposed by the Australian Commonwealth Scientific and Industrial Research Organization for the treatment of aluminous laterites by acid pressure leaching with recovery of basic aluminium sulphate by hydrolysis. However the latter requires a high temperature for calcination and the resulting alumina is sintered and unsuitable for feeding to the electrolytic reduction cell.

The question of minimum economic size for a hydrometallurgical zinc plant was also discussed and the view was expressed that plants smaller than 50,000 tons production per annum would probably not be viable.

REFERENCES

1. Stevens, L. G., Goeller, L. A. and Miller, M.: The U.O.P. Nickel Extraction Process. 14th Annual Conference of Metallurgists, Edmonton, Canada, August 1975.
2. Siemens, R.E., Good, P.C., and Stickney, W.A.: Recovery of Nickel and Cobalt from Low Grade Domestic Ores. U.S. Bureau of Mines, 1975 Report of Investigations No. 8027.
3. Subramanian, K.N., and Jennings, P.H.: Review of the Hydrometallurgy of Chalcopyrite Concentrates. Canadian Metallurgical Quarterly, 1972, 11, No. 2, pp. 387–400.
4. Dyson, N.F., and Scott, T.R.: Acid Leaching of Nickel Sulphide Concentrates. Hydrometallurgy, Vol. 1 (in press).
5. Biangardi, S., and Pietsch, H.: Treatment of Sulphidic Copper Ores by the L.M. Process. Institution of Chemical Engineers, Symposium Series No. 42, April 1975, Paper 5.
6. Roords, H.J., and Queneau, P.E.: Recovery of Nickel and Cobalt from Limonites by Aqueous Chlorination. Trans. Inst. Mining and Metallurgy, June 1973, 82 C. 79–87.
7. Iammartino, N.R.: Uranium Processing Update. Chem. Eng.: March 3 1975, pp. 88–89.
8. Zubryckyj, N., Evans, D.J.I., and Mackiw, V.N.: Preferential Sulphation of Nickel and Cobalt in Lateritic Ores. J. Metals, May 1965, 17, pp. 478–86.
9. Scott, T.R.: Continuous Co-current Pressure Leaching of Zinc—Lead Concentrates under Acid Conditions. A.I.M.E. International Symposium on Hydrometallurgy, Chicago, 1973, pp. 718–750.
10. Boldt, J.R.: The Winning of Nickel. 1967, Methuen (London).
11. Bush, P.D., Engle, L.F., Gates, E.H., and Vijayaraghavan, M.D.: The Pressure Leaching-Cementation-In-Pulp Process for Nickel Laterites and Sulphides. A.I.M.E. International Symposium on Hydrometallurgy, Chicago, 1973, pp. 63–92.
12. Scott, T.R.: Alumina by Acid Extraction. J. Metals, Feb. 1962, pp. 121–125.
13. Cathro, K.J.: The Recovery of Copper from Chalcopyrite Concentrates by Means of Sulphur Activation, Cupric Chloride Leaching and Electrolysis. Dec. 1974, Proc. Australasian Institute of Mining and Metallurgy 252, pp. 1–11.
14. Prater, J.D., Queneau, P.B., and Hudson, T.J.: A Nitric Acid Route to Processing Copper Concentrates. A.I.M.E. Annual Meeting, San Francisco, 1972, Preprint B-91.
15. Caron, M.H.: Ammonia Leaching of Nickel and Cobalt Ores. J. Metals, Jan. 1950, 2, pp. 67–90.
16. Anon.: Development of the Inco Iron Ore Recovery Process. Canadian Mining and Metallurgical Bull. 1956, 49, pp. 337–343.
17. Koch, W.E., and Pietsch, H.B.: Treatment of Oxide Copper Ores by Reduction and Ammoniacal Leaching. Trans. Institution of Mining and Metallurgy, March 1974. 83, C14–17.
18. Kuhn, M.C., Arbiter, N., and Kling, H.: Anaconda's Arbiter Process for Copper. C.I.M. Bulletin, Feb. 1974, 67, pp. 62–73.
19. Bussell, M., Ewers, W.E., Thornber, M.R., Dyson, N.F., and Scott, T.R.: The Use of Pyrrhotite to Recover Nickel and Cobalt from Acid Leach Liquors. A.I.M.E. International Symposium on Hydrometallurgy, Chicago, 1973, pp. 1125–1143.
20. Thornhill, P.G., Wigstol, E. and Van Weert, G.: The Falconbridge Matte Leach Process. J. Metals, July 1971, pp. 13–18.

21. Fonseca, A.G.: Ammonia-Oxidative Leach of Chalcopyrite. C.I.M. Bulletin, July 1974, 67, pp. 105–110.
22. Evans, D.J.I.: The Production of Metals by Gaseous Reduction from Solution. Advances in Extractive Metallurgy Institution of Mining and Metallurgy, April 1967, pp. 831–907.
23. Palperi, M., and Aaltonen, O.: Fluid Bed Sulphatizing Roasting and Leaching at the Outokumpu Oy Cobalt Plant. A.I.M.E. Annual Meeting, Denver, 1970, Preprint A-55.
24. Chaimungalanon, S., and Robins, R.G.: Precipitation of Metallic Copper by Carbohydrate Reduction of Sulphate Solutions. Australian Chemical Engineering, Nov. 1973, 14, pp. 9–15.
25. Bitzer, E.C.: Cementation-In-Pulp of Copper Leach Liquors. Engineering and Mining Journal, October 1974, 175, pp. 72–78.
26. Skewes, H.R.: Electrowinning of Lead Directly from Galena, Proceedings, Australasian Institute of Mining and Metallurgy, 1972, 244, pp. 35–41.
27. Biegler, T., and Constable, D.C.: Upgrading and Activation of Chalcopyrite Concentrates by Slurry Electrolysis. Trans. Institution of Mining and Metallurgy, 1975 (in press).
28. Eliasen, R.D., and Edmunds, E.: The S.E.C. Nickel Process. C.I.M. Bulletin, Feb. 1974, 67, pp. 82–85.
29. Fleet, D.S.: Solvent Extraction in Copper Hydrometallurgy: a Review. Trans. Institution of Mining and Metallurgy, March 1974, 83, C 30–38.
30. Allen E.S., and Kruesi, P.R.: Cymet Electrometallurgical Process for Treating Base Metal Sulphide Concentrates. Joint Meeting M.M.I.J.-A.I.M.E., Tokyo, 1972, Print No. T.iv.b.4.
31. Ritcey, G.M., and Lucas, B.H.: Extraction and Separation of Copper, Nickel, Zinc and Cobalt from Ammoniacal Solution using Kelex 100. C.I.M. Bulletin, Feb. 1975, 68, pp. 105–113.
32. Robinson, R.E., and Velthuis, R.G.: The History and Development of the Ion Exchange Process as applied to Uranium Extraction. South African Institute of Mining and Metallurgy, Jan. 1957, 57, pp. 376–385.
33. Slater, M.J.: Continuous Counter-Current Ion Exchange in the Mining Industry. C.I.M. Bulletin, Feb. 1974, 67, pp. 93–98.
34. Warshawsky, A.: Solvent Impregnated Resins in Hydrometallurgical Applications. Trans. Institution of Mining and Metallurgy, June 1974, 83, C 101–104.
35. Bolto, B.A., Dixon, D.R., Eldridge, R.J., Swinton, E.A., Weiss, D.E., Willis, D., Battaerd, H.A.J. and Young, P.H.: Use of Magnetic Polymers in Water Treatment. Journal of Polymer Science Symposium 1975, 49, pp. 211–219.
36. Ross, J.R., Salisbury, H.B., and Potter, G.M.: Pressure Stripping Gold from Activated Carbon. A.I.M.E. Annual Meeting, Chicago, 1973. (Reprint available from U.S. Bureau of Mines, Salt Lake City, Utah, U.S.A.).
37. Gordon, A.R., and Pickering, R.W.: Improved Leaching Technologies in the Electrolytic Zinc Industry. Metallurgical Transactions B. March 1975, 6, pp. 43–53.
38. Davey, P.T., and Scott, T.R.: Removal of Iron from Leach Liquors by the Geothite Process. Hydrometallurgy Vol. 1 (in press).
39. Matthew, I.G., and Elsner, D.: Hydrometallurgical Treatment of Siliceous Zinc Ores. Australian Patent No. 407,500, 13 May 1968.
40. Anon.: Marinduque's Nonoc Nickel Mine and Plant. World Mining, October 1971, 24, pp. 66–68.
41. Whitehead, A.B., and Urie, R.W.: Fluidized Bed Roasting of Copper in Concentrates. Proc. Australasian Institute of Mining and Metallurgy, 1961, pp. 61–85.
42. Bes de Berc, O.: New Aluminium Production Methods. Aluminium. 1975, 51 (6), pp. 380–382.
43. Wilmshurst, R.E., and Parker, R.L.: The Burra Treatment Process. Australasian Institute of Mining and Metallurgy, Adelaide Regional Conference, August 1971. Paper No. 10.
44. Griffith, W.A., Day, H.E., Jordan, T.S., and Nyman, V.C.: Development of the Roast-Leach-Electrowin Process for Lakeshore. A.I.M.E. Annual Meeting, Chicago, 1973, Preprint No. 64: J. Metals, Feb. 1975, 27, pp. 17–25.
45. Van Weert, G., Mah, K., and Piret, N.L.: Hydrochloric Acid Leaching of Nickeliferous Pyrrhotites from the Sudbury District. C.I.M. Bulletin, Jan. 1974, 67, pp. 97–103.

46. Schmittroth, W.: The Removal of Magnesia and Lime from Dolomitic Type Zinc Concentrate. A.I.M.E. Annual Meeting, New York, 1958 (Preprint).
47. Paynter, J.C.: A Review of Copper Hydrometallurgy. Journal of the South African Institute of Mining and Metallurgy, Nov. 1973, 74, pp. 158–170.
48. Sitnai, O., and Peeler, J.P.K.: Evaluation of Four Potential Hydrometallurgical Processes for Copper Production. Proc. Australasian Institute of Mining and Metallurgy, 1975. (in press).
49. Dasher, J.: New Copper Extraction Processes. (Panel Discussion) J. Metals, August 1974, 26 pp. 23–24.
50. Minerals Yearbook, 1972: United States Department of the Interior, U.S. Government Printing Office, Washington, 1974, pages 1234–1235.
51. Rupay, C.H., and Jewell, C.J.: Regeneration of Hydrochloric Acid from Waste Pickel Liquor. C.I.M. Bulletin, Feb. 1975, 68, pp. 89–93.
52. Jacobi, J.S.: Industrial Cell Design and Operation. Chemistry and Industry, 3 May 1975, pp. 370–374.

PAPER 5

IN SITU EXTRACTION AND LEACHING TECHNOLOGY

by Roshan B. Bhappu

IN SITU, IN-PLACE, chemical or solution mining is the in-place extraction of metals from ores located within the confines of a mine (unfractured or fractured ore, stope fill, caved material, and ores in permeable zones) or in dumps, ore heaps, slag piles, and tailing ponds. These materials represent an enormous, untapped, potential source of all types of metals. The field of *in situ* mining, encompasses the preparation of ore for subsequent in-place leaching, the flow of solutions and ionic species through rock masses and within rock pores, the leaching of minerals with inexpensive and regenerable leaching reagents under conditions prevailing in-place, the generation and regeneration of such solutions, and the recovery of metals or metal compounds from the metal-bearing solutions. Accordingly, the overall scope of this potential mining method embraces interdisciplinary science and technology requiring application of the principles of basic sciences, mineral technology, hydrology, and economics.

Heretofore, this kind of mining has been more or less limited to the extraction of copper from low-grade materials, however, it has a much greater potential than this. Practically all metals are susceptible to leaching in the *in situ* environment. Processes will soon be developed for the in-place extraction and recovery of metals such as lead, zinc, nickel, manganese, uranium, silver, gold, molybdenum, and mercury.

Advantages

1. *In situ* mining is environmentally more attractive, since it creates less surface disturbance and results in less water and air pollution than conventional mining.

2. *In situ* mining usually requires less capital and lower operating cost than a conventional mine and milling operation. Moreover, the total energy required in solution mining is much lower than that required to obtain the same quantity of metal by conventional processing methods.

3. *In situ* mining can often be used to recover metals economically from materials that could not be so treated by more conventional mining, milling, and smelting processes.

4. *In situ* mining increases a mine's ore sources and reserves. Low-grade or inaccessible ore zones, dumps and tailings may become ores, thereby leading to more effective mineral utilisation and conservation.

5. The leach solutions obtained through *in situ* mining usually lend them-

selves to a variety of metal recovery processes. The pure metal or metal compounds so obtained may be of greater value to a mine owner than the sulphide or oxide products normally obtained by conventional milling processes.

6. *In situ* mining can often be used in conjunction with a conventional mining or milling process to boost metal recoveries and increase ore reserves.

Disadvantages

1. Both physical and chemical restrictions may limit the usefulness of *in situ* mining. The effectiveness of contacting ore with solutions and the recovery of leach solutions from the system without appreciable loss are two important physical factors. Dissolution rates, metal precipitation, and solution regeneration chemistry are major factors.

2. Testing an *in situ* mining situation short of actual field operation sometimes proves difficult.

3. Ground water contamination may result from *in situ* mining operations.

4. Much of the data on the physical, chemical, and bacteriological factors involved in *in situ* extraction has not been evaluated commercially.

Technological Considerations

In Situ Leaching Techniques

There are basically three types of leaching systems in the field of *in situ* leaching, depending mainly on the physical location of the ore deposit. These situations are illustrated in Figure 1.

Typical examples of Type 1 are the numerous dump leaching operations practiced by the open pit copper mines in the Southwestern United States. *Dump leaching* is used in openpit mining operations to dispose of lower grade (submarginal) stripped material in a remote area with the intention of recovering the valuable metals contained therein. Because of limited available area and easy placement, such dumps tend to be very high and the leaching of such piles, either through spraying or ponding of solution, is relatively slow and inefficient. Nevertheless, the production of copper from such dump leaching operations at the present amounts to about 200,000 tons per year in the United States or approximately 15% of the total production. It should be noted that the effectiveness of copper extraction in these dump leaching operations has been partially attributed to bacterial oxidation of sulphide minerals. Figure 2 schematically illustrates various components of a typical dump leaching operation.

Several successful heap leaching operations for extraction of copper, gold, and uranium as carried out in USA (Arizona, Nevada, Texas, New Mexico, and Wyoming) also belong to Type 1. *Heap leaching* is a system in which low grade or mixed oxide-sulphide ores are leached in relatively small heaps on

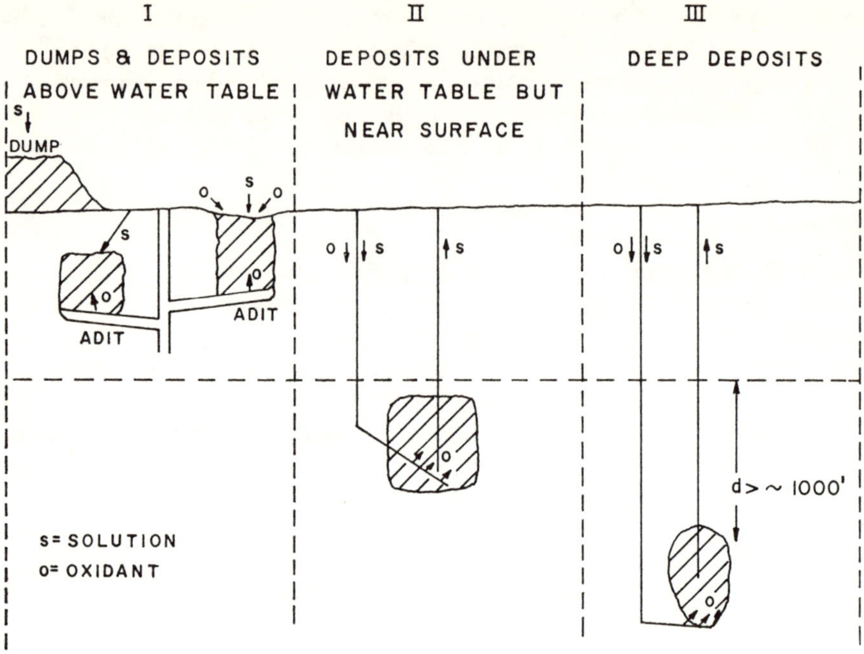

Figure 1—Three types of in situ leaching situations. (according to Dr. Milton Wardsworth)

prepared pads or surfaces with positive drainage over a shorter leaching period under controlled conditions. In some cases new heaps are piled on top of other heaps and in other cases the ore is brought to a designated site for treatment and removal, followed by piling of fresh ore on the same site for treatment.

Other viable in-place leaching situations belonging to Type 1 are the truly *in situ leaching* systems in which the ore is broken in-place by some type of blasting technique followed by percolation leaching of the fragmented ore mass. Appropriate collection galleries are provided to collect the leach solution for subsequent recovery of metal and recirculation of the barren solution for additional leaching. Since the ore is broken and left in place, such leaching systems are economically attractive. Ranchers Exploration and Development Corporation's Old Reliable Mine Operation in Arizona (Figure 3) and the Big Mike Mine in Nevada (Figure 4) are fine examples of this *in situ* leaching method.

Also, belonging to Type 1 are the classic *in situ* leaching operations of the Miami and Ray Mines, Arizona, USA, in which the leaching has been carried out on the fractured submarginal ore zones resulting from block caving operations.

In the case of the Miami Mine operation, it has been estimated that 85,000 tons of copper metal have been produced over a 15-year period since com-

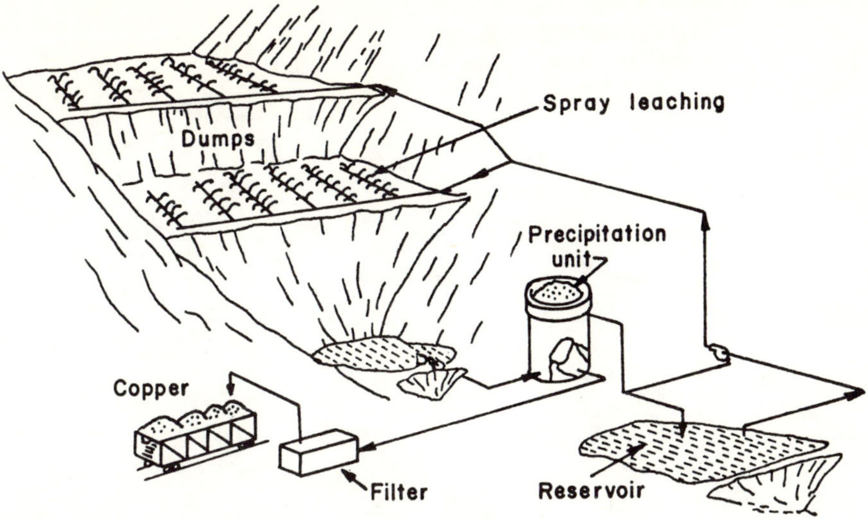

LEACHING
$CuFS_2 + 2Fe_2(SO_4)_3 \longrightarrow CuSO_4 + 5FeSO_4 + 2S$
FERRIC SULPHATE GENERATION
$4FeSO_4 + O_2 + 2H_2SO_4 \xrightarrow{\text{bacteria}} 2Fe_2(SO_4)_3 + 2H_2O$
IRON REMOVAL BY HYDROLYSIS
$2Fe^{+2} + O_2 + 2H^+ \longrightarrow 2Fe^{+3} + H_2O$ (slow)
$6Fe^{+3} + 4SO_4^{-2} + 14H_2O \longrightarrow 3Fe_2O_3 \cdot 4SO_3 \cdot 9H_2O + 1OH^+$
CEMENTATION WITH SCRAP IRON
$CuSO_4 + Fe \longrightarrow Cu + FeSO_4$
SOLVENT EXTRACTION
$2RH + Cu^{+2} \longrightarrow R_2Cu + 2H^+$
ELECTROWINNING
Cu^{+2}

Figure 2—Basic components of copper dump leaching

pletion of the underground mining operation in 1959, with an estimated recovery of about 60–70% of the metal left in place. The leaching operation continues with no decrease in recovery rate and no immediate end in sight.

The second (Type II) situation reflects the leaching of relatively shallow deposits located less than 300 metres from the surface and which are under the water table. Such deposits may need to be fractured in-place and the solution drained in order to commence the alternate leaching and drying cycles. On the other hand, metal values from such deposits may be extracted by utilising bore hole mining technology as currently practiced by uranium mining operations in the United States. The *Bore Hole Mining* technique usually consists of a properly designed pattern of wells suitable for injection of the leaching solution and recovery of metal-bearing liquors (Figure 5). The wells are like water wells except that the casings are perforated for optimum leaching conditions.

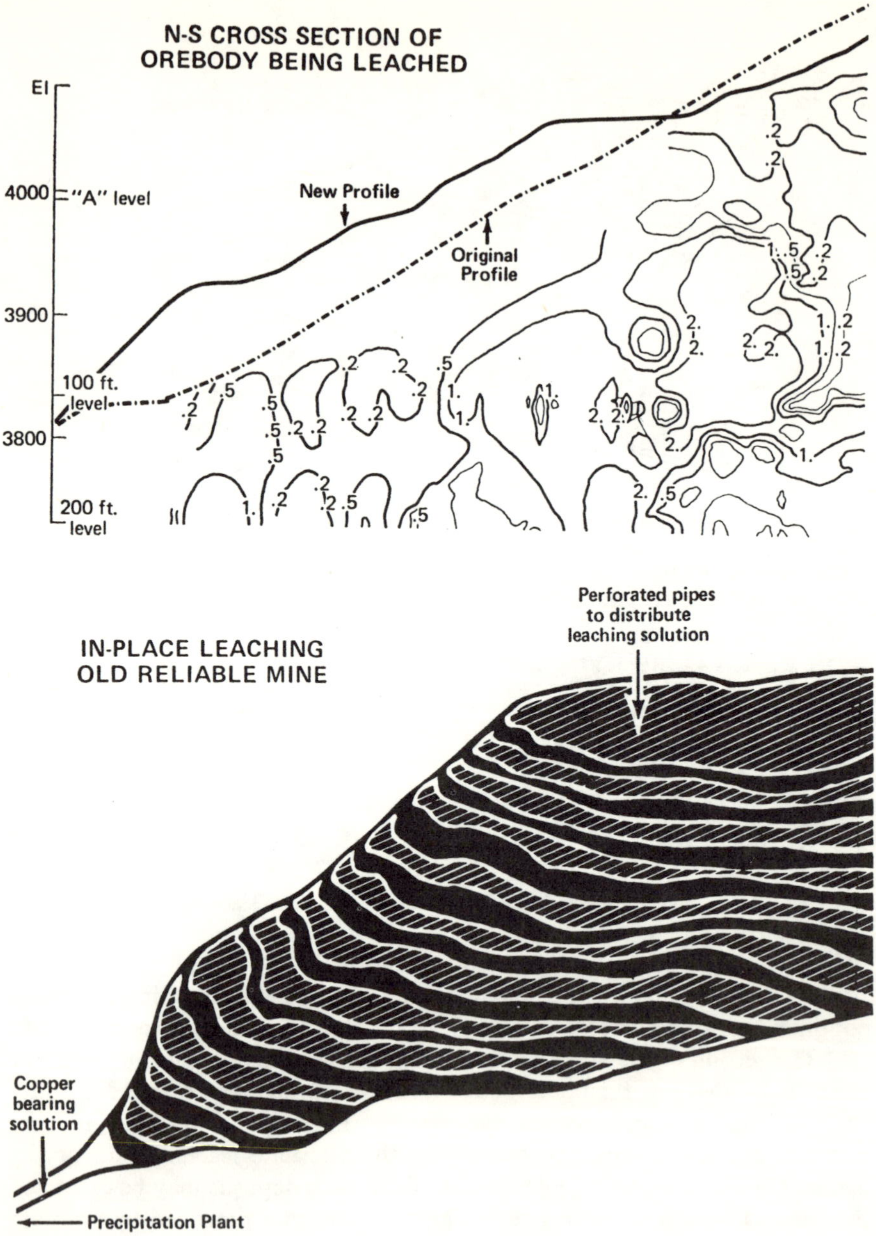

Figure 3—Cross section of Old Reliable Mine Ore body and in-situ leaching after blasting and contouring.
(E.I. du Pont de Nemours & Co. Inc.)

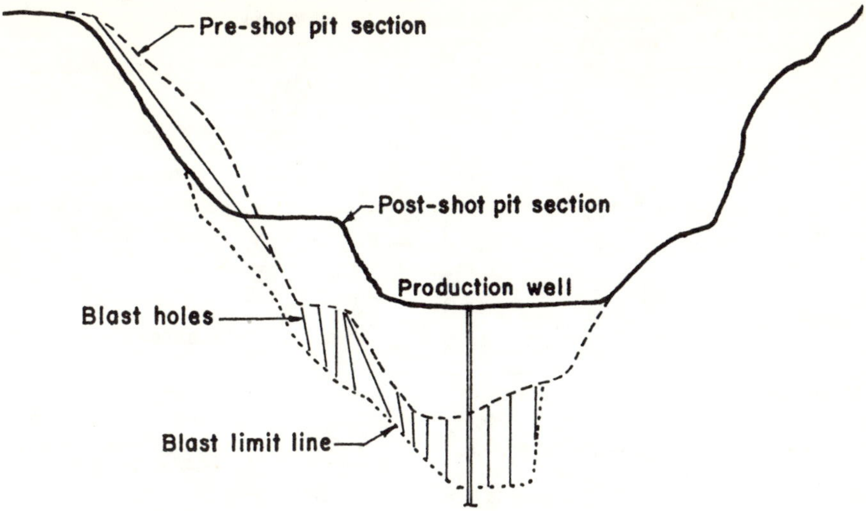

Pit cross-section before and after blast

Figure 4.—In situ leaching of copper at Big Mike Operation, Nevada (Ranchers Exploration & Development Corporation)

Type III situations are characterised by relatively deep-seated deposits in excess of 300 metres and under the water table. Leaching of such deposits through chemical mining has been proposed by Lawrence Livermore Laboratory, University of California, USA, after fracturing the ore body by conventional or nuclear devices (Figure 6). Leaching at such depths is enhanced

Table 1: In Situ Leaching Practices in USA

Dump Leaching
1. Several operations in S.W. USA (copper)
Heap Leaching
1. Blue Bird Mine, Arizona (copper) 2. Inspiration Operations, Arizona (copper) 3. Johnson Camp Mine, Arizona (copper) 4. San Juan Mine, Arizona (copper) 5. Carlin and Cortez Gold Operations, Nevada (gold) 6. Round Mt., Nevada (gold)
In Situ Leaching (Copper)
1. Miami, Arizona (block caving operation) 2. Ray, Arizona (block caving operation) 3. Mt. City, Nevada (block caving operation) 4. Old Reliable, Arizona (4 million lb. AN-FO coyote blast) 5. Big Mike, Nevada (400,000 lb. AN-FO pit wall blast) 6. Zonia, Arizona (4.14, 0.9, and 1.5 million lb. AN-FO blasts) 7. Nuclear Blast (AEC-Kennecott concept for deep-seated deposits) 8. Kennecott's Experiment (near Safford, Arizona)
In Situ Leaching (Uranium)
1. Utah Mining and Construction Company, Shirley Basin, Wyoming 2. Mine Water Leaching at Grants, New Mexico 3. Bacterial Leaching at Elliot Lake, Canada 4. Wyoming Minerals, (NH_3 LEACH), Bruni, Texas 5. Wyoming Minerals, Ray Point, Texas 6. Arco-US Steel-Dalco Operation, George West, Texas 7. Union Carbide, Duval County, Texas 8. Exxon Operations, Powder River Basin, Wyoming 9. Mobil Oil Corporation, Webb County, Texas

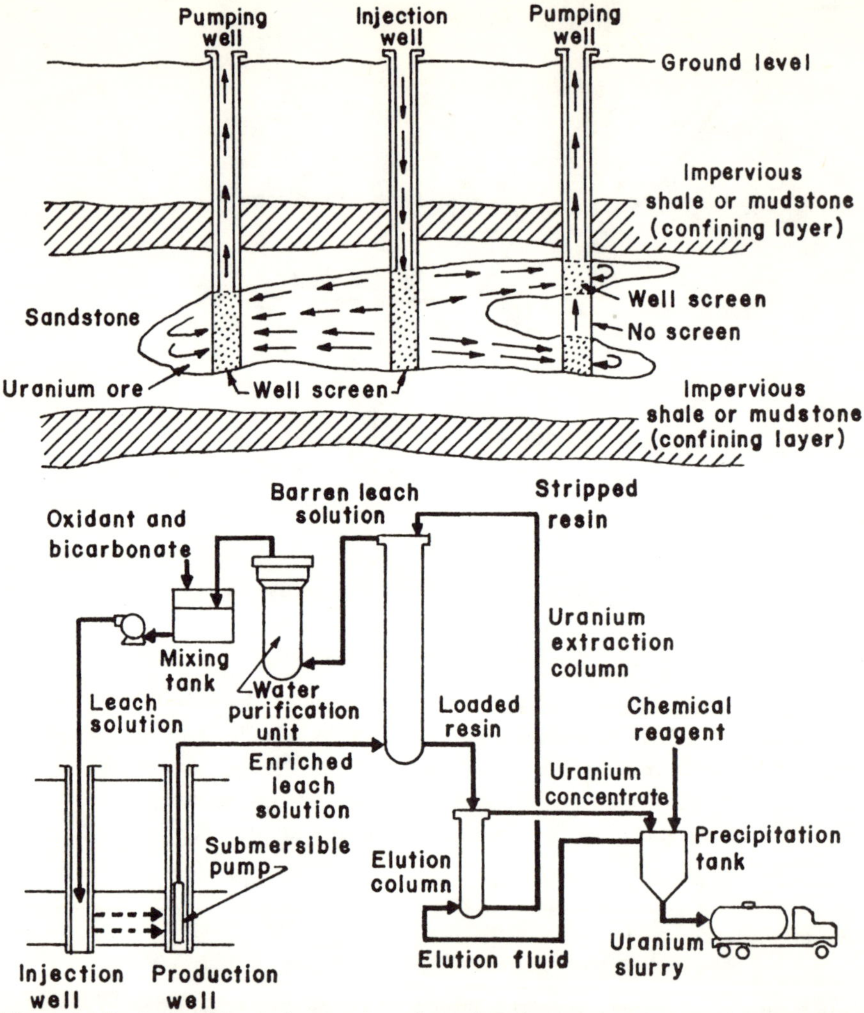

Figure 5—In situ leaching and extraction of uranium through bore hole mining technique. (according to E.E. Malouf, Kennecott Copper Corporation)

by direct oxidation of sulphide minerals due to increased solubility of oxygen resulting from hydrostatic head.

A listing of various USA commercial and pilot plant operations employing the above three types of in-place leaching techniques in the extraction of copper, gold, and uranium is given in Table I. These examples of practical and economic operations clearly indicate that *in situ* extraction technology has developed into a matured and viable metal extraction process.

Controlling Parameters in In Situ Leaching

Successful *in situ* leaching is governed by several factors including

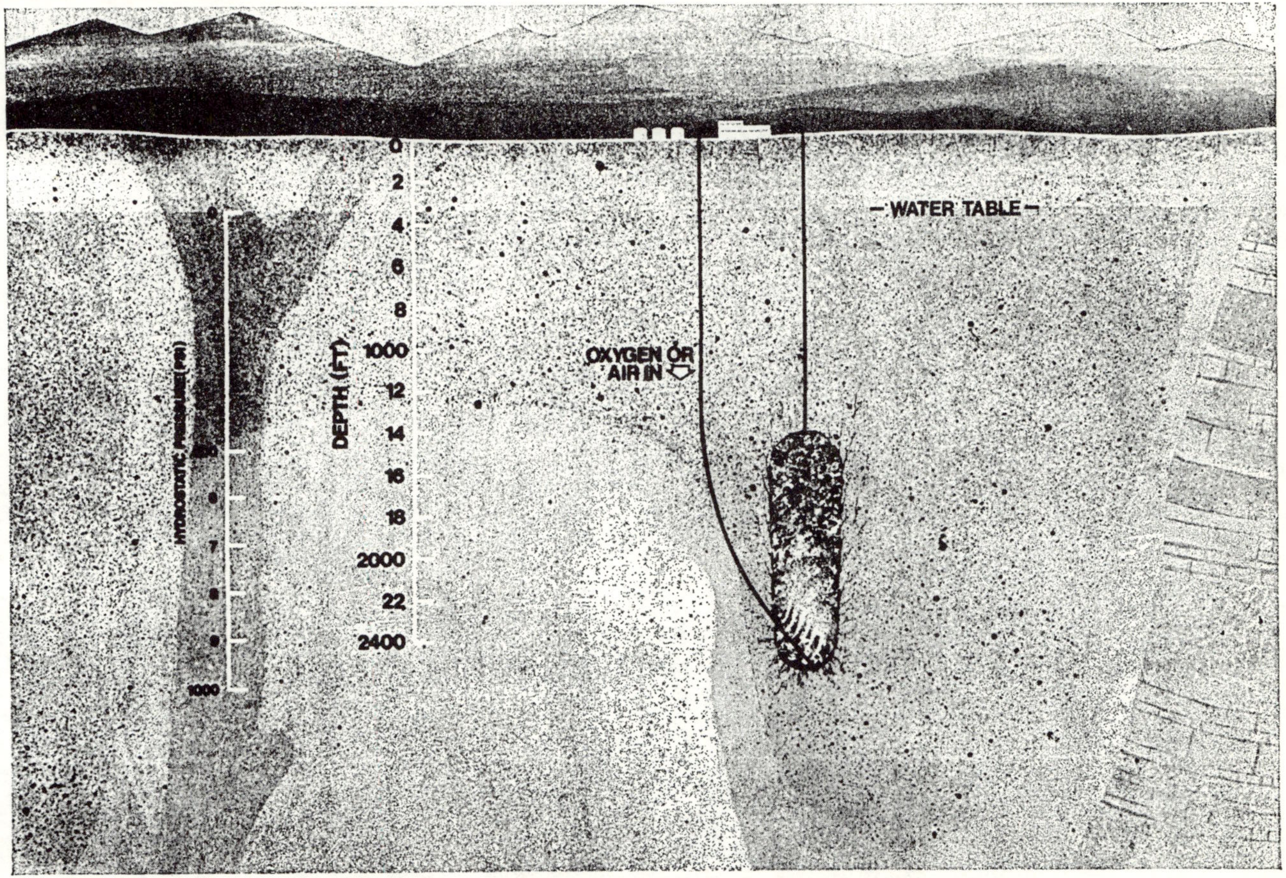

Figure 6—Conceptual illustration of in situ leaching after nuclear fracturing (Lawrence Livermore Laboratory, University of California)

geological and mineralogical considerations, elements of the leaching phase, solubility of minerals in various leaching reagents, effect of particle size, effective fragmentation, solution application and regeneration, and metal recovery methods.

Role of Mineralogy

Broadly speaking, the chemical processing techniques used for dissolving minerals *in situ* are the reverse of the ore-genesis processes which resulted in the emplacement of the minerals. A better understanding of the chemistry and physics of emplacement will in due course lead to the elaboration of more precise methods of recovery. It is important, in geological evaluations, not to exclude low-grade and complex mineral deposits from ore-reserves if their mineralogic characteristics indicate possible use of *in situ* mining technology. It is equally important in metallurgical investigations to consider the chemical and physical nature of the processes of deposition of the ore minerals in relation to extraction possibilities.

The most significant impact of this relationship, however, is in *in situ* mining and in-place leaching for the recovery of metals from low-grade and deep-seated deposits. Because this process involves the dissolution of metals and minerals under geochemical and geophysical conditions in an underground environment, reversing the mechanisms associated with ore deposition and mineral formation would be of utmost importance in putting the metal values in their ionic form to be easily recovered by simple solution flow and pumping. Such processes are already available for recovering copper and uranium values from submarginal ores. The technique can be extended to the recovery of lead, zinc, nickel, molybdenum, vanadium, tungsten, and other important metals.

It is interesting to note that the effectiveness of *in situ* leaching of copper, uranium, and gold by appropriate reagents is economically feasible because of favourable mineralisation of the metal values in the host rock. This is especially true for successful leaching of oxidised gold ores at coarser size since these ores contain submicron size gold particles in fracture fillings and cavities in the host rock consisting of either limey or silicified siltstones, sandy dolomite, and schist. These host rocks also tend to be porous and remain permeable throughout the leaching cycle. In the case of oxide-copper ores, the mineralisation also occurs along macro and micro fractures in the host rock, while in the case of uranium, the values are in the form of secondary-enriched precipitates coating the sand grains of the sandstone host.

Elements of the Leaching Phase

Accessibility, physico-chemical interaction, and transport constitute the elements of the leaching phase involved in *in situ* extractions. Limitations imposed on any of these factors restrict the leaching process.

Accessibility is essential because interaction between the desired constituents and the lixiviant cannot take place in the absence of contacts, which depend on exposure and penetrability. The factors to consider are locations of the values, their volume and shape distribution, exposure area, specific surface, particle size, porosity, capillary pressure, viscosity pressure, solubility of gases in leach solutions, and surface roughness.

Physico-chemical interaction converts the desired constituents from a fixed to a mobile condition and is governed by the solubility of the solid in leach solutions, vapour pressure in gases, and the presence of bacteria. Knowledge of free energies of reactants and products helps to determine whether a reaction is possible. The kinetic factors involved include time, concentration, diffusivity, specific rate constants, wettability, and oxidising conditions, especially due to bacteria activity.

The first two elements by themselves do not ensure successful leaching without transport of products away from and reactants to the zone of interaction through diffusion and convection. Diffusion is governed by concentration gradient and diffusivity, which in turn are influenced by particle size, micropore radius, temperature, and molecular mass. On the other hand, convective flow concerns interparticle penetration and is restricted by pressure gradient, permeability, viscosity, and surface roughness.

In general, therefore, the factors governing leaching can be grouped into three major classifications: namely; physical, chemical, and biological. The majority of the leaching studies in the past have emphasised chemical factors. However, it is essential that we consider the physical factors also, since they definitely influence the leaching process. We must develop new techniques for physical and chemical testing of ore samples and for establishing the limitations of and determining optimum parameters for successful extraction of values from the broken ore. Moreover, in recent years, considerable interest has been shown in the beneficial effect of bacterial activity in oxidation and thus the dissolution of copper and uranium under *in situ* leaching environments. In such cases, oxidising bacteria of the *Thiobacillus thiooxidans* and *Thiobacillus ferrooxidans* groups have been identified as the active agents. Realising the importance of these bacteria in leaching systems considerable effort is made, in practice, to encourage their presence and growth.

Solubility of Minerals

In general, nearly all the minerals are more or less soluble in either acids, alkalies, or other specific reagents such as cyanide.

Most of the oxide-copper minerals such as azurite, malachite and chrysocolla are quite soluble in sulphuric acid. For this reason, sulphuric acid is the preferred leaching reagent for *in situ* extraction of oxide-copper ores. However, in cases when the oxide-copper ore contains considerable amounts of calcite, the acid consumption is very high and it may not be economical to use sulphuric acid for leaching. In this case a caustic or an ammonia leach would be preferable. Thus, the solubility of gangue may have a profound influence in the selection of the appropriate reagent for leaching of valuable minerals.

As far as the sulphide copper minerals are concerned, simple sulphuric acid leaching is not very effective and the presence of ferric sulphate is necessary to dissolve minerals such as chalcocite under *in situ* leaching conditions. The strong oxidising conditions required for chalcocite dissolution are satisfied by the presence of ferric iron. Many commercial operations have failed because the required oxidising conditions were not maintained during the leaching process. This is also the reason why the presence of the oxidising bacteria *Thiobacillus thiooxidans* is necessary in dump, heap and *in situ* leaching operations. The required iron is either available in the natural ore or is added to the leach system through the cementation step in which copper in solution is precipitated as elemental copper when the liquid is contacted with scrap iron.

Chalcopyrite also requires oxidising conditions for its dissolution and under typical dump, heap, and *in situ* leaching conditions not much oxygen is available. For this reason, deposits containing chalcopyrite at or near the surface are not effectively leached and it is common for only 10 to 25% of the copper to be recovered over many years of leaching operation, the practical success of the dump leaching operations being attributed to the presence of huge amounts of copper contained in the dumps and correspondingly high output.

Chalcopyrite can, however, be effectively leached under autoclave conditions of high temperatures (80 to 90°C) and pressures (400–500 psi) which are normally found under *in situ* conditions encountered in the leaching of deep-seated copper deposits. For this reason, recovery of copper from deep-seated primary copper deposits is economically feasible.

In the case of in-place leaching of uranium, both dilute sulphuric acid and carbonate-bicarbonate solutions, are effective reagents for dissolving uranium minerals. However, both the reagents require a chemical oxidant to oxidise uranium from the insoluble oxidation state to the soluble state. Typical acid leach oxidants are chlorates or manganese dioxide and typical carbonate leach oxidants are hydrogen peroxide or hypochlorites.

For gold and silver leaching, cyanide has been found to be the most selective reagent, and is the universal solvent used. Here again, oxygen is required in the reaction and efforts must be made to maintain oxidising conditions during the in-place leaching of gold.

Effect of Particle Size and Fragmentation

Since leaching of minerals by various chemical reagents involves effective

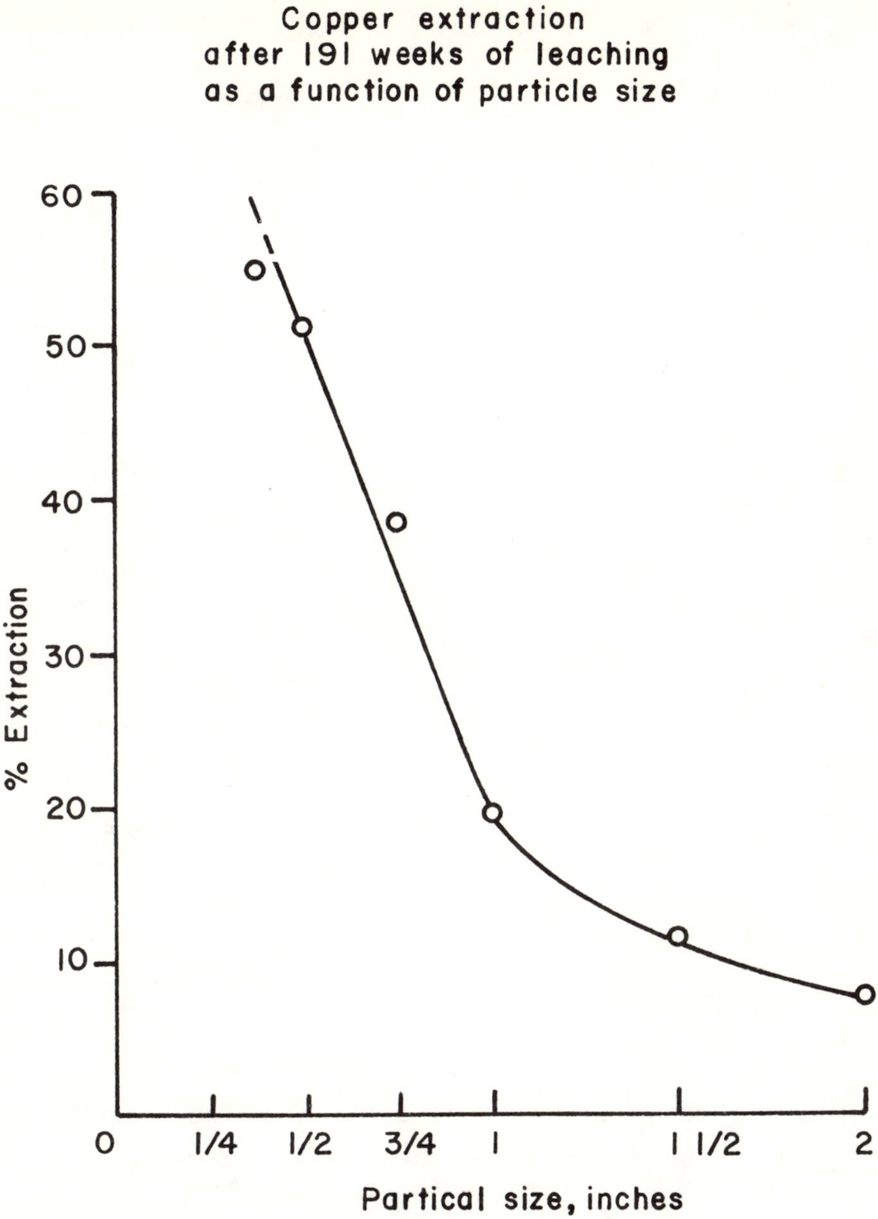

Figure 7—Effect of particle size on leaching of copper.

contact between the two, it is obvious that the mineral must be liberated or freed from the host rock. The only other way the reagent can contact the valuable mineral is by effective penetration or diffusion through fractures, cracks, capillary action, and other physical factors considered under the

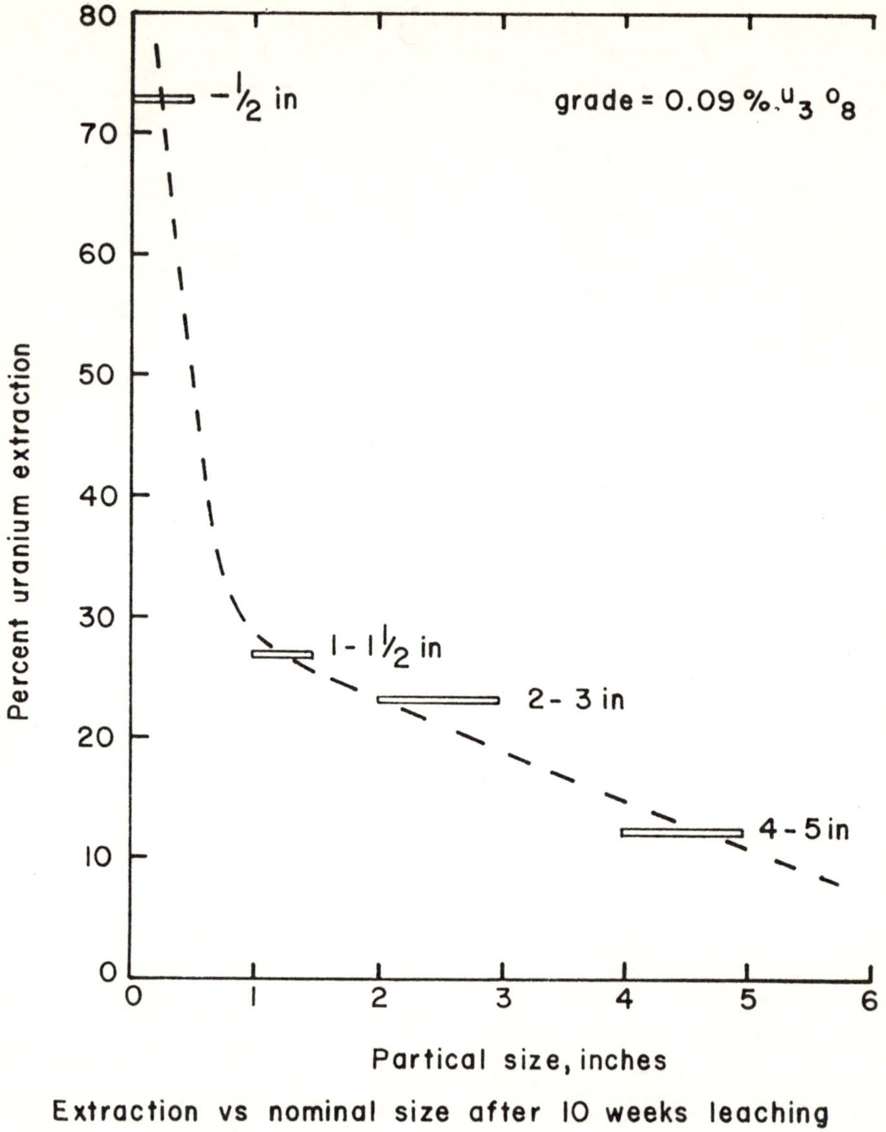

Figure 8—Effect of particle size on leaching of uranium.

"elements of the leaching phase". Regardless of the process of contact, it is quite evident that the reduction in particle size would be of considerable benefit in *in situ* leaching of minerals. With finer ore size it would be possible to obtain higher extraction rates in shorter leaching periods.

The beneficial effect of finer particle size is clearly indicated for copper, uranium, and gold ores in Figures 7, 8 and 9 respectively.

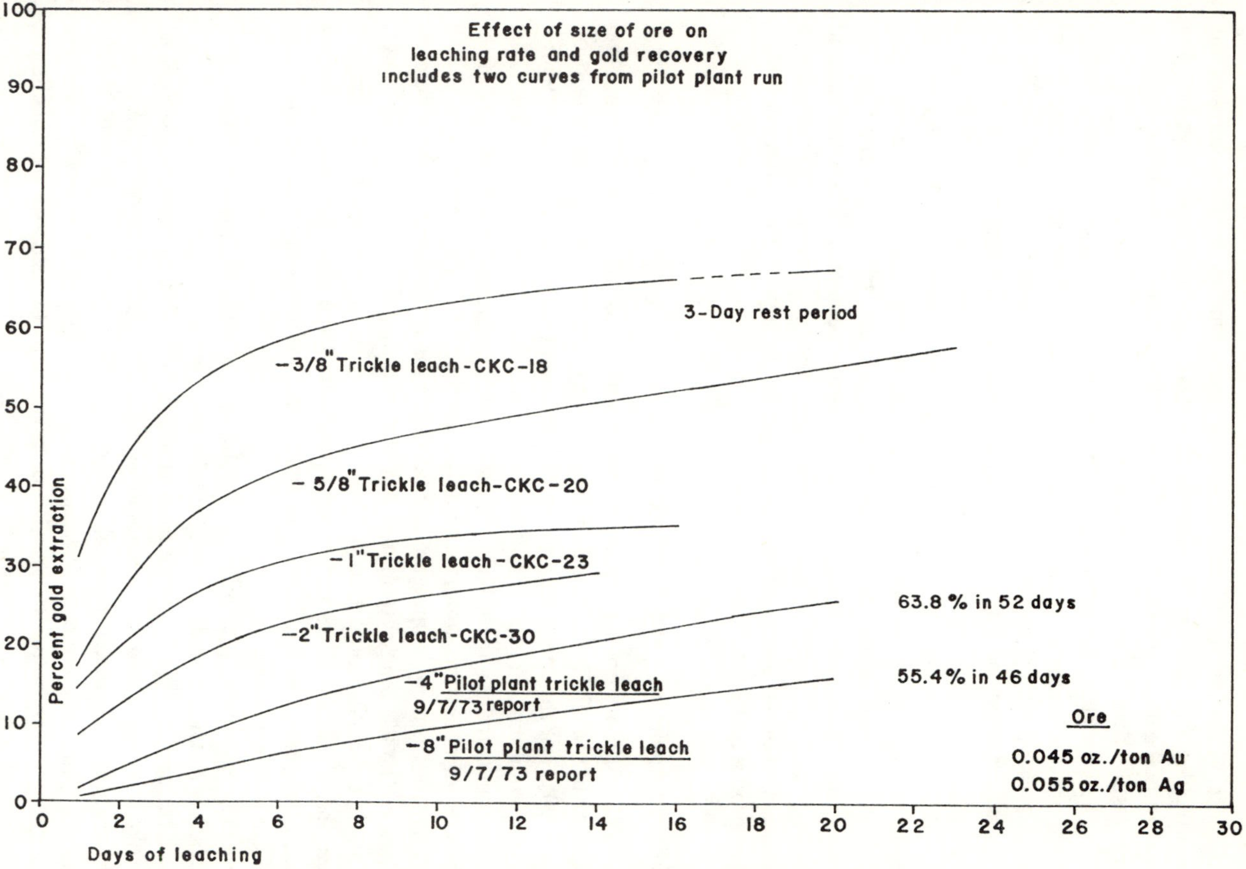

Figure 9—Effect of particle size on leaching of gold.

The desired particle size in *in situ* leaching can be obtained through proper blasting and fracturing techniques. The blast design for *in situ* leaching closely follows the principles of blast design used in more conventional mining practices. However, in the case of *in situ* mining unusually large dimensions are involved and these pose inherent problems associated with large blasting situations. To date, several unusually large blasts have been made for leaching with results that are very encouraging.

The important factors that need to be considered in engineering a blast for *in situ* leaching are (a) shape of deposit, (b) characteristics of ore body, (c) seismic activity encountered, (d) missile consideration, (e) type of explosive, (f) powder factor and distribution, (g) initiating and delay system, (h) loading logistics, (i) economics, and (j) safety.

Besides conventional fracturing by blasting compounds, efforts have been made to explore the possibility of using liquid explosives, hydrofracturing and nuclear explosives. No doubt these and many other new concepts for fracturing will be considered as *in situ* leaching becomes a more conventional method in the future. Figure 10 illustrates the fragmentation encountered with conventional blasting practice in a porphyry copper body compared with that produced by nuclear explosions.

Solution Application and Regeneration

The primary goals in *in situ* leaching are: (a) to recover maximum metal content in the minimum solution volume and (b) to recover the major portion (plus 90%) of the specified volume of leach solution fed to the system with a given minimum amount of metal content over the life of the operation. This minimum content should be such that the value of recovered metal will offset the cost of operation, amortisation and overhead and yield a profit. It is also important from environmental considerations that the leach solution be confined in the leaching system and not report as a pollutant in ground water. This is especially important in gold leaching using cyanide as the specific reagent. Both the toxic nature of cyanide and the value of leach solution containing gold necessitates *in situ* leaching conditions approaching a "zero discharge" objective.

The two principle systems of solution application: (a) downward percolation or trickle leach and (b) totally submerged system.

In the first, the leach solution is usually distributed over the top of the broken ore by ponding or spraying, with the solution flowing by gravity through the fractured material to the collection system. Such a technique allows optimum contact as well as oxidising conditions necessary for effective leaching. Leach solution can also be applied to broken ore by injecting it through drill holes containing perforated casing.

In the second system, the leach solution is injected into the consolidated or unbroken ore zone located under the ground water table and is forced

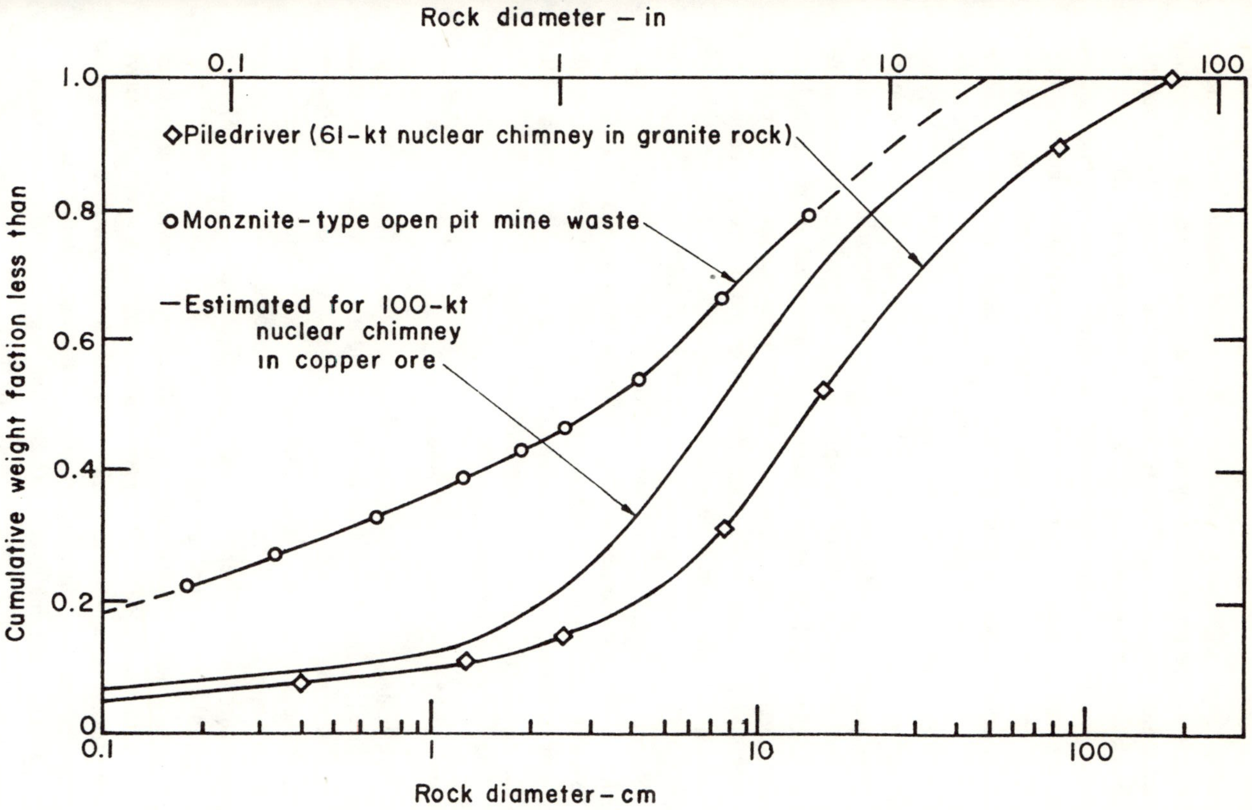

Rock size distribution

Figure 10—Comparison of rock size distributions resulting from different fracturing practices. (Lawrence Livermore Laboratory, University of California)

through the permeable ore body to dissolve the metal. The metal-containing solution is then recovered through appropriate recovery-wells. Thus a series of injection, recovery, and monitoring wells are used. This system is used for uranium and copper through the so called "bore hole mining" technique.

Inasmuch as the chemical reagents used in *in situ* leaching generally constitute a major cost item and greatly influence the economics of leaching, reagent generation and regeneration play a very important part in any in-place leaching process. Some reagents can be generated and regenerated by natural processes in the leaching cycle, whereas others require chemical processing. The only known example of autoregeneration of reagent is the natural reaction involved in the production of sulphuric acid and ferric sulphate from pyrites and spent ferrous sulphate solution under natural aeration or biological oxidation conditions. Most of the other reagents are generated and regenerated by various industrial methods such as the contact process for sulphuric acid, electrolysis for generating hypochlorite and autoclaving to obtain ferric sulphate. In some cases, part of the reagent is regenerated by the metal extraction process incorporated in the flowsheet, such as the regeneration of sulphuric acid through solvent extraction and electrowinning as practised in the extraction of copper from leach solutions. In other cases, though the regeneration of reagent from spent solutions is possible, it is not economically feasible or environmentally attractive, such as the regeneration of cyanide from spent solution in gold leaching.

The key to low-cost reagents for *in situ* leaching probably lies in the efficient use of either low-cost materials found in the ore to be leached, like pyrite, or raw or low-cost prepared chemical products readily available, like pyrite, sodium chloride, trona or ammonia.

The Recovery of Metals from Leach Solution

The last phase of any hydrometallurgical process, including *in situ* leaching, is the recovery of metals from leach liquors. Conventional purification of a metal-containing solution followed by recovery of metals or compounds from the solution by either chemical or electrolytic precipitation is employed to obtain the marketable product. These recovery techniques are adequately covered in the literature and their effectiveness is clearly demonstrated in several successful plant practices.

In connexion with *in situ* leaching applications, however, the recovery phase poses certain technical problems that may influence the overall effectiveness of the process. One such difficulty concerns treating a large volume of very dilute metal-bearing solution. This may require recirculation of the leach solution to build up the metal content and then bleeding off of a small part of the concentrated leach stream for metal recovery.

Newer techniques of ion exchange, solvent extraction, and charcoal adsorption are being used for effective concentration of metals from leach

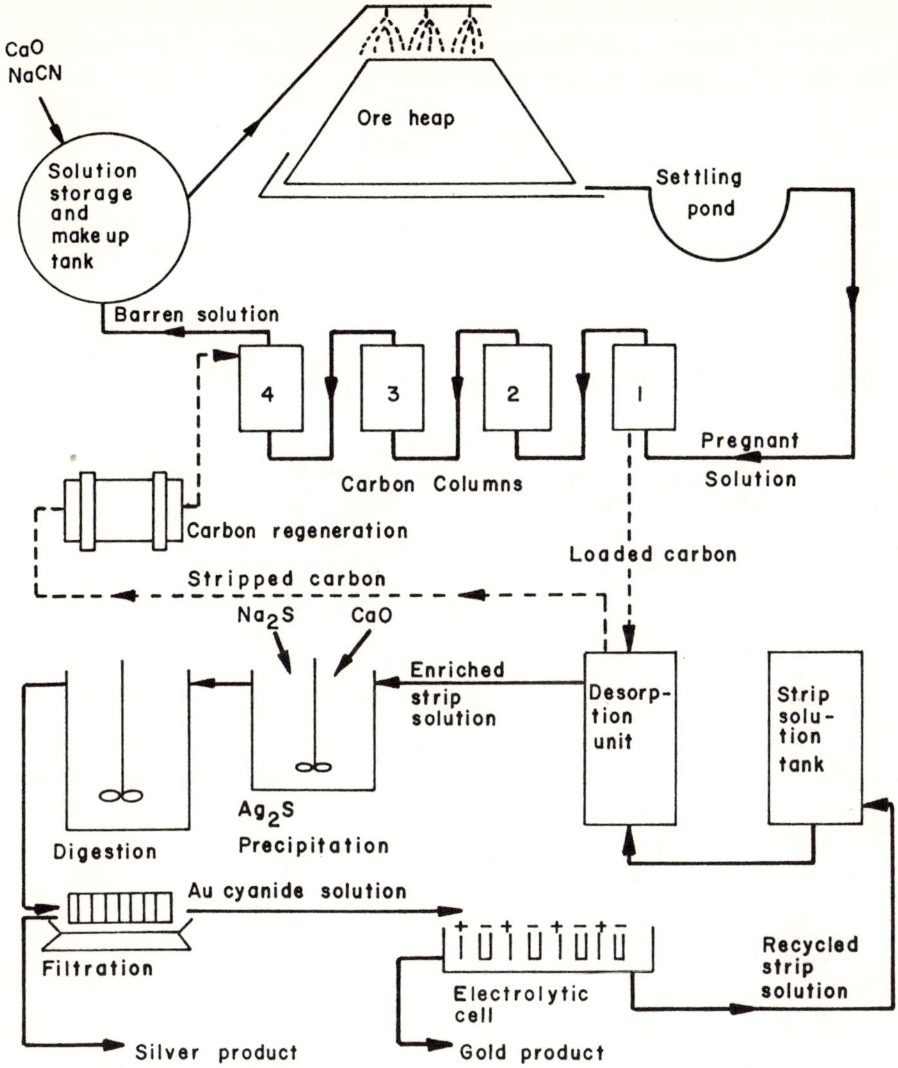

Figure 11—Proposed flowsheet for heap leaching of gold ores (U.S.B.M.)

solutions. Moreover, these procedures have proved very effective for processing large volumes of leach solutions containing more than one valuable metal. In recent years, the recovery of copper from leach solutions by solvent extraction and electrowinning has become more attractive than the copper cementation process used conventionally all over the world. Similarly, ion exchange has been exclusively used for recovery of low concentrations of uranium from *in situ* leaching operations. Finally, the use of activated carbon for the recovery of gold and silver values (as low as one part per million or less) from in-place leaching (Figure 11) operations has become a favourite process. In this case, the gold is subsequently extracted by desorbing gold values from carbon by hot caustic solution followed by electrolysis to obtain the gold.

Research Needs

Although *in situ* leaching in recent years has matured into a viable metal extraction process, considerable additional research is needed to bring it into competition with other high-confidence and universally acceptable processes. The principal areas of research required to accomplish the above goal are:

1. Improved techniques for measuring the physical properties of ores in-place.
2. Improved techniques for drilling and blasting to produce desired fragmentation.
3. Improvement in injection and recovery systems.
4. Improved understanding of hydrology in broken ore masses.
5. Design of leaching systems for maximum oxidation.
6. Basic studies on leaching of dispersed mineral particles.
7. Improved methods of scaling up from laboratory tests.
8. Optimum parameters for leaching of massive deep-seated deposits.
9. Improved environmental controls.

Systematic Plan for In Situ Leaching Operation

Experience in the field of *in situ* leaching for the extraction of copper, uranium and gold has indicated the following systematic approach for initiation of *in situ* leaching ventures:

1. Exploration—geology—mineralogy
2. Sampling
3. Testing: rock mechanics and leaching
4. Preliminary feasibility study
5. Environmental evaluation
6. Seismic testing
7. Hydrological studies
8. Drilling and development
9. Solution recovery design
10. Optimisation of leaching phase
11. Final feasibility study
12. Fracturing
13. Leaching
14. Solution collection
15. Metal recovery
16. Modifications
17. Modelling
18. Optimisation

Following is an outline of pertinent data required for a feasibility study, modified from the one proposed by W. J. Robinson, Anaconda Company, USA.

I. Property

A. Location, descriptions, acreage, maps

1. Mine 2. Plant site 3. Tailings and ore waste disposal 4. Townsite

B. Legalities

1. Ownership 2. Mining claims 3. Fee land 4. Timber and water rights 5. Rights of way 6. Use permit

C. Access

D. Climate

E. Environmental Considerations

1. Air pollution 2. Water pollution 3. Health hazards 4. Effect on neighbouring areas

F. Acquisition plans and costs

II. Geology

A. Development

1. Churn drilling (a) Spacing (b) Footage (c) Caving and effect on samples
2. Diamond drilling (a) Spacing (b) Footage (c) Core recovery
3. Rotary drilling (a) Spacing (b) Footage (c) Wet or dry sampling

B. Type of deposit

C. Structure

D. Mineralisation

E. Ore reserves—(mine planning)

1. Types of ore or minerals 2. Grades 3. Tonnages 4. By-products 5. Complete chemical analysis of ore and concentrates 6. Pre-production stripping 7. Discard ratio (a) Overburden (b) Included waste (c) Sub-marginal minerals

III. Treatment Process

A. Selection of process based on ore reserves and grades (flotation, hydrometallurgical, or *in situ* leaching)

B. Leaching methods considered

1. Dump 2. Heap 3. Vat 4. *In situ* 5. Combinations

C. Recovery plants suggested

1. Cementation 2. Ion exchange—solvent extraction—electrowinning 3. Carbon adsorption—desorption—electrowinning

D. Bench scale tests

1. Source of samples 2. Metallurgical tests and balances 3. Other results 4. Need for pilot plant 5. Complete analysis of finished project

E. Pilot plant

1. Size 2. Source of samples 3. Metallurgical tests and balances 4. Other results 5. Complete analysis of finished product

F. Flowsheet

1. Crushing (if required) 2. Leach system 3. Recovery plant

G. Plant

1. Tons per day 2. Equipment (a) Type and quantity (b) Capital costs (c) Operating costs 3. Manpower (a) Number and qualifications (b) Costs 4. Operating supplies and maintenance costs 5. Other costs (a) Buildings (b) Metal credits (c) Transportation (d) General

IV. Mine Planning

A. Leaching requirements

B. Mining method if required

C. Tons per day

1. Stripping ratio

D. Equipment requirements

1. Type and quantity 2. Capital costs 3. Operating costs

E. Manpower requirements

1. Number and qualifications 2. Costs

F. Operating supplies and maintenance costs

G. Other costs

1. Buildings 2. Taxes 3. Royalties 4. Transportation 5. General

V. Ancillary Facilities

A. Water supply

1. Source 2. Ownership—water rights 3. Quality and quantity 4. System (a) Type (b) Cost

B. Power supply

1. Source 2. Costs

C. Fuel supply—gas, oil, coal

1. Source 2. Costs

D. Shops and garages

E. Warehouses

F. Change house

G. Offices

H. Health and welfare

I. Other

VI. Labour Supply

A. Requirements

B. Availability

C. Costs

D. Labour relations in area

VII. Housing

A. Requirements

B. Availability

C. Plans to meet requirement

1. Company houses 2. Private housing development

VIII. Transportation
A. Railroad
B. Highway
C. Water
D. Other

IX. Management
A. Corporate level
B. Local level—organisation chart

X. Refining or other treatment
A. Where
B. Costs

XI, Marketing
A. Market survey
B. Prices
C. Agent

XII. Financial Outcome Analysis
A. Pre-production time and costs
B. Year-by-year plans
1. Expenditures 2. Productions 3. Income 4. Cash flows
C. Present worth or yield
D. Variances and sensitivity

XIII. Completed Feasibility Report

Economic Considerations

Feasibility Studies

From a well organised and realistic feasibility study such as that described in the previous section the profit potential and the viability of the project can be assessed by all the parties concerned with the development of the mine. Such a feasibility study also assists in screening and evaluating various choices and options involved in mining and processing. This is especially true for the development of lower grade deposits, since the paramount question here is the best method of extracting the metal values.

In the case of *in situ* leaching, especially, the feasibility study plays a very significant role since the percentage recovery of metal is very difficult to predict. Even though the laboratory or pilot plant tests indicate a certain recovery, say 60 or 70%, there is no guarantee that in actual practice it would be possible to attain the specified recoveries. In such cases it behoves the mine management to carry out the feasibility study and cash flow analyses at 60, 50, 40, 30, 20, and 10% recovery levels. If the cash flow and thus the economic viability of the project does not appear to be favourable for 50 or

40% recovery levels, then it would be preferable to cancel the project or undertake additional laboratory and field studies in order to increase the confidence level of the economically acceptable recovery figure. On the other hand, if the feasibility study appears favourable for 30 or 20% recovery levels, all efforts should be made to initiate the *in situ* leaching operation as soon as possible. Recent *in situ* leaching operations at Old Reliable Mine, Arizona and Big Mike Mine, Nevada (by Ranchers Exploration and Development Company), have both proved economically successful at 15 to 20% overall copper extraction.

Since the cash flow and thus the overall economic feasibility of an *in situ* leaching venture may be influenced significantly by the price of metal, it is customary to carry out feasibility studies at selected metal prices below and above the prevailing price.

Capital and Operating Cost Estimates

Table 2 shows the estimated capital costs for unit processes encountered in typical *in situ* leaching operations for extraction of specified quantities of copper, gold and uranium. It should be noted that in some cases the capital cost is based on quantity of metal produced per day while for others, it is based on the volume of leach solutions processed. Using these figures it is possible to estimate the capital cost of similar processes for other production schedules (pounds or ounces per day) and leach solution volumes (gpm).

Table 3 shows estimated operating costs for various unit operations encountered in *in situ* and heap-leaching extraction of copper, gold, and uranium. As can be seen, the cost of solvent extraction—electrowinning of copper is considerably less than for cementation. The primary reason for this difference is the additional cost of smelting and refining (\$0.20 to \$0.28 per lb) required for cemented copper which is not required for high purity cathode copper produced in the solvent extraction—electrowinning process.

Table 2: Estimated Capital Costs for Unit Processes

UNIT PROCESS	*BASIS*	*ESTIMATED COST*
Copper (40,000 lbs/d; 4,000 gpm at 1.0 gpl Cu)		
Cementation	\$70,000–80,000/t Cu/d	\$1,400,000 to 1,600,000
Solvent Extraction	\$150–200/ft.^{2} settling area	\$3,000,000 to 4,000,000
Electrowinning	\$60 to 70/lb cathode Cu	\$2,400,000 to 2,800,000
Total SX-EW	\$135 to 170/lb Cu/d	\$5,400,000 to 6,800,000
Gold (200 oz/d; 2,000 gpm; 0.016 oz Au/t solution)		
Carbon Adsorption	\$250–300/gpm	\$500,000 to 600,000
Desorption, Electrowinning	\$1,750–1,800/oz Au/d	\$350,000 to 360,000
Total ADS—DES—EW		\$850,000 to 1,010,000
Uranium (2,750 lbs/d; 2,000 gpm at 0.125 gpl U_3O_8)		
Ion Exchange	\$1,750–2,000/gpm	\$3,500,000 to 4,000,000
Precipitation	\$450–500/lb U_3O_8/d	\$1,237,500 to 1,375,000
Total IX-PPT		\$4,737,500 to 5,375,000

Table 3: Estimated Direct Operating Costs for In-Place Extraction

UNIT OPERATION	*ESTIMATED COST*
Fracturing in place (with development)	\$0.30 to 0.40/t
Mining (drilling, blasting, hauling and dumping)	0.40 to 0.50/t
Crushing (primary, secondary, tertiary)	0.11–0.22–0.33/t
COPPER ORE AVERAGING 0.5 *TO* 1.0% *Cu*	
Leaching (heap; pads; 4.0 lbs H_2SO_4/lb Cu)	\$0.35 to 0.45/t
Leaching (*in situ*, 4.0 lbs H_2SO_4/lb Cu)	0.22 to 0.32/t
Cementation	0.10 to 0.12/lb
Smelting, refining, and transportation	0.20 to 0.28/lb
Solvent extraction—electrowinning	0.07 to 0.09/lb
Cost/lb Cu (*in situ*—cementation)	0.39 to 0.46
Cost/lb Cu (*in situ*–SX—EW)	0.15 to 0.22
Cost/lb Cu (heap-cementation)	0.50 to 0.66
Cost/lb Cu (heap—SX—EW)	0.26 to 0.42
GOLD ORE AVERAGING 0.03 *TO* 0.06 *oz/t*	
Leaching (*in situ*; 1.0 lbs. NaCN; 3.0 lbs., CaO)	\$0.45 to 0.55/t
Leaching (heap with pads; 1.5 lbs NaCN; 3.0 lbs., CaO)	0.55 to 0.65/t
Carbon adsorption, desorption & electrowinning	0.16 to 0.20/t
Cost/t (*in situ*—carbon adsorption—EW)	1.48 to 1.58
Cost/t (heap with pads—carbon adsorption—EW)	1.79 to 1.89
URANIUM ORE AVERAGING 0.05 *TO* 0.10% U_3O_8	
Bore hole drilling (with casing)	\$13.00 to 18.00/ft
Leaching (drilling and leaching-carbonate leach)	2.00 to 3.00/lb
Cost/lb U_3O_8 (*in situ* leach—IX—PPT)	5.00 to 7.00

It is also interesting to note that in the processing of gold ores either by *in situ* or heap leaching, the overall operating costs vary from \$1.48 to \$1.89 per ton. These figures indicate that using these leaching techniques it may be profitable to extract gold values from ores averaging 0.03 to 0.05 ounce per ton when the price of gold is above \$100 per ounce.

Finally, in the case of extracting uranium by "bore hole mining" technology, the overall operating costs vary from \$5.00 to \$7.00 per pound U_3O_8 (yellow cake) produced. Since the current price of uranium in the USA is in excess of \$25 per pound U_3O_8, it may be possible to treat uranium ores containing grades as low as 0.02 per cent U_3O_8.

Case Histories

In order to demonstrate the economic viability of the *in situ* leaching systems and to demonstrate the usefulness of preliminary feasibility studies, two case histories, one of *in situ* leaching of copper and the other pertaining to bore hole mining of uranium, will be reviewed.

(1) *In Situ Leaching for Copper*

The hydrometallurgical processes currently available for treating oxide-copper ore include dump leaching, heap leaching, vat leaching, agitation leaching, and *in situ* leaching in combination with the conventional cementation (with scrap iron) or the newly developed solvent extraction-electrowinning system for copper production.

From the viewpoint of metallurgical effectiveness, in general, dump and heap leaching give lower extraction of copper amounting to 40 to 60% in several weeks to months, depending on the ore size and height of the dump or heap. On the other hand, agitation leaching gives the highest extraction of copper, amounting to about 90% in four to twenty-four hours, followed by vat leaching with an extraction of about 80% in six to eight-day leaching cycles. However, agitation and vat leaching both require high capital costs and are more expensive to operate. For this reason, they are applicable only to deposits having huge ore reserves and grades in excess of 1% copper. *In situ* leaching is comparatively slow and may provide 50 to 60% extraction of copper in a variable leaching period, depending on the nature of the deposit and the type of operation used. But since *in situ* leaching requires

Table 4: Cost Comparison (0.66% *Cu* = 8.0 *lbs/ton recovered*) *Direct Operating Costs Per Ton Ore*

UNIT OPERATIONS	*CONVENTIONAL* (strip ratio = 1:1)	*IN SITU*
Drilling		
Blasting		
Hauling	\$0.70–0.90	\$0.35–0.45
Dumping		
Crushing	0.15–0.30	
Ore Handling	0.10–0.15	
Leaching (with pad)	0.40–0.50	0.20–0.30
Metal Recovery (SX-EW)	0.56–0.64	0.56–0.64
Supervision and Administration	0.35–0.40	0.25–0.30
Total	\$2.26–2.89	\$1.36–1.69
Cost/lb	\$0.28–0.36	\$0.17–0.21

Table 5: Preliminary Feasibility Study In Situ Leaching for Copper

Ore reserve	5,000,000 tons	
Grade	0.66% Cu	
Mine Life	7 years at 715,000 tpy	
Estimated recovery	60% or 8.0 lbs/ton	
Daily productions	16,000 lbs	
CAPITAL COST		
Leaching facilities	\$200,000	
SX-EW (with LIX inventory)	3,000,000	
Supporting facilities	100,000	
Cost for processing		\$3,300,000
Cost for fracturing (5 million tons at 0.40/t)		2,000,000
Total capital cost		\$5,300,000
Operating cost for processing		\$1.12 per ton

lower capital cost and is cheaper to operate, it can be applicable to deposits that have limited tonnage and lower grade. Table 4 shows the comparative operating cost of fracturing and leaching of a copper ore assaying 0.66% copper by conventional against *in situ* mining.

In the case history illustrated here, we are concerned with a 5 million ton oxide-copper deposit with a grade of 0.66% copper. The *in situ* leaching system employed will consist of blasting ore in-place and leaching the broken ore with sulphuric acid under *in situ* environment followed by recovery of copper from solution by solvent extraction and electrowinning. The leach system and recovery plant are designed to produce 16,000 pounds of copper per day over a 7 year mine life.

The capital and operating costs as shown in Table 5 are taken from the pertinent data provided in previous tables. Table 6 shows the preliminary cash

Table 6: Preliminary Cash Flow Analysis (In Thousands of Dollars)

Capital (3,300,000+2,000,000)	$5,300
Sales (5,760,000 lbs at $60.0)	3,456
Operating (715,000 at $1.12/ton)	800
Depletion (15% of sales)*	518
Depreciation (straight line)	757
Local tax and insurance (8% of sales)	276
Total operating cost	2,353
Operating income	1,103
Income tax (48%)	530
Net Operating income	574
Investment credit (10% capital or 1/2 net income)*	287
Net income	860
Working capital (20% of operating cost)	471
Cash flow	1,665
Payout time	3.2 years
ROI (discontinued to present worth)	30%

* USA situations

flow analysis for this particular *in situ* leaching operation. As can be seen, the payout time (time required to pay back the capital expenditure) is about 2–3 years with a return of investment (ROI = average interest rate, discounted to present worth, to return the investment in full) of 30%. Obviously, this study indicates an economically attractive venture.

(2) *Bore Hole Mining for Uranium*

The *in situ* mining of lower grade uranium ores has received increasing attention in recent years due to the favourable price of uranium in the free market. An in-place extraction technique of considerable promise is the so-called "bore hole" mining in which the uranium is recovered by drilling into the ore body, circulating a lixiviant fluid to dissolve the mineral, extracting the uranium values from the pregnant solution, regenerating and recycling the lixiviant. Such a technique is economically and environmentally attractive in extracting uranium values from deeper, lower grade reserves.

In the evaluation of a block of uranium-mineralised zone as a potential producer by in-place leaching using the bore hole mining technique, it is first necessary to establish a grade-thickness product which will cover development costs and operating costs with something left over for recovery of investment in plant and equipment and profit. This grade-thickness product is then a cutoff parameter for determination of whether another hole is included in reserves or excluded.

To calculate the grade-thickness product of a hole, it is necessary to establish a second cutoff, the grade at which operating costs are just covered. Amortisation of development and plant and equipment costs are not included in calculating this grade. No sample interval of a grade less than this operating breakeven grade can be included in calculation of the grade-thickness product of the hole for determination of the potential contribution of the hole to reserves.

Once the holes have been classified as a potential ore or waste, the reserves in each potential block can be estimated, for example, by the ore outline method.

Development costs for determination of whether a hole is or is not a potential contributor to ore include:

1. Costs of drilling and casing injection and production wells
2. Pumps
3. Surface piping and electrical requirements to bring an ore block into production

Operating costs in the field include:

1. Operating and repair costs of injection of solvent and of reagents required to convert uranium minerals to soluble form.
2. Operating and repair costs of pumping the pregnant solution from collecting point to the ion exchange plant and costs for recovering the uranium minerals.

Now an injection and production well pattern can be finalised and any necessary adjustments to development costs can be made. The capital cost for the piping to carry the solution from the ore blocks to the ion exchange plant, the capital cost of the ion exchange and precipitation plant, and other ancillary equipment can be estimated.

A cash flow analysis is now made to determine whether the project has satisfactory profit potential. At this stage, it is necessary to be sure that a realistic estimate of the true thickness of strata which will be penetrated by solution has been made. Large amounts of reagents will often be consumed in reactions in sub-ore grade zones above and below the "ore interval". Also, there will often be one or more sub-ore zones within an ore interval which will also consume reagents.

As a general principle, it should be anticipated that even after careful

Table 7: In Situ Uranium Leaching Operation

Type of Operation	Bore Hole Mining
Deposit	30-foot Ore Zone at 400-foot depth
Reserves	4,000,000 ton deposit at 0.1% U_3O_8
Contained U_3O_8	8,000,000 pounds
Recoverable U_3O_8	5,600,000 pounds at 70% Recovery
Plant Capacity	1,000,000 pounds U_3O_8 per year 2,750 pounds per day
Production life	5.6 Years
Type Process	Ion exchange-precipitation
Type Leaching	Acid at 100 pounds H_2SO_4 per ton of ore
Leaching Rate	2,000 gpm
Solution Grade	0.125 gpl U_3O_8
Capital Cost, Leaching	$3,400,000
Capital Cost, IX-PPT	$5,000,000 (IX = $4,000,000; PPT = $1,000,000)
Total Capital Cost	$8,400,000
Operating Cost, Leaching	$3.00 per pound U_3O_8
Operating Cost, IX-PPT	$4.00 per pound U_3O_8
Total Operating Cost	$7.00 per pound U_3O_8

Cut-off Grade Based on:
Operating Cost at $8.00 per pound 0.062% U_3O_8
Operating Cost at $16.00 per pound 0.032% U_3O_8
Operating Cost at $24.00 per pound 0.020% U_3O_8

Table 8: In Situ Uranium Leaching Economics (Bore Hole Mining—IX—PPT)

CASH FLOW (In Thousands of Dollars)

	Price of U_3O_8			
	$8/*lb*	$16/*lb*		$24/*lb*
Capital Cost	$8,400	$8,400	$8,400	$8,400
Net Sales (at 70% Extraction)	8,000		16,000	24,000
(at 35% Extraction)		8,000		
Costs and Expenses:				
Total Operating	7,000	7,000	7,000	7,000
Depreciation (1)	1,500	1,500	1,500	1,500
Depletion (2)	1,760	1,760	3,520	5,280
Indirect Cost (3)	350	350	350	350
Total	10,610	10,610	12,370	14,130
Operating Income	<1,210>	<1,210>	3,630	9,870
Income Tax (4)	—	—	1,742	4,738
Net Operating Income	<1,210>	<1,210>	1,888	5,132
Investment Credit (5)	—	—	840	840
Net Income	<1,210>	<1,210>	2,728	5,972
Cash Flow	2,050	2,050	7,748	12,752
Pay-Out in Years	4.1	4.1	1.1	0.7
Return on Investment (6)	12%	12%	86%	117%

Notes:
(1) Depreciation—Straight-Line, 5.6 years
(2) Depletion—22% of Net Sales (USA situations)
(3) Indirect Cost—5% of Total Operating Cost
(4) Income Tax—48%
(5) Investment Credit—10% of Capital Cost for the First Year, not to exceed 50% of Net Operating Income in One Year. (USA situations)
(6) Rate of Return Discounted to Present Worth

geological and engineering studies after months of actual experience in a field, it will still probably not be possible to predict leaching rates or total recovery from one block to the next with anything approaching accuracy. For example, one block may yield 120% of the estimated recoverable uranium in three to six months. The adjacent block, although apparently geologically the same, may never yield over 60% of its estimated recoverable uranium regardless of the number of months it is leached. It is necessary, therefore, to be fairly conservative in estimating the pounds of U_3O_8 which must be present in a block to justify leaching.

Table 7 gives the details of the capital and operating costs for the second case-history involving the feasibility of extracting uranium values from a given sandstone deposit. Table 8 shows the financial analysis for the project under consideration.

As can be noted, the bore hole mining technique appears to be a very attractive method for extracting uranium from lower grade ores. At the old price of $8.00 per pound U_3O_8 the project is not economically favourable. However, at a uranium price of over $16.00 per pound, the project appears to be very attractive with ROI in excess of 85%.

Discussion Summary

In discussion, it was estimated that the capital cost of *in situ* leaching plant in Latin America would be 50% greater than those in the United States and those in Africa 100% to 300% greater depending on availability of infrastructure. It was anticipated that 1980 costs would be about double those of 1975. Reference was made to the fact that *in situ* treatments were not labour-intensive (as normally regarded as desirable in developing countries); to the effect of humid and rainy climates on dump-leaching operations (requiring adjustment of some operations); to the effectiveness of the methods for small-tonnage situations; to the control of cyanide solutions and of the silica content of solutions; to the quantity of lixiviant required (0.004 to 0.1 gallons per minute per square foot) and to the method of application (trickle or spray); to the recovery of uranium from lignite ash; to the low cost of the cementation of copper in Chile (5–6 cents per lb) and to the performance of bacteria in leaching (poor in deep-seated deposits and where oxygen lacking). The minimum ore-grades for *in situ* treatments were assessed as follows: Copper: Dump leaching, 0.16% Cu; blasting and heap leaching, 0.35%–0.40% Cu or down to 0.25% Cu if transport costs negligible (Cu at 66 cts/lb).

IN SITU LEACHING BIBLIOGRAPHY—PART I (by USBM)

1. Anderson, A.E., and F.K. Cameron: Recovery of Copper by Leaching. Ohio Copper Company of Utah. Trans. AIME, V. 73, 1926, pp. 31–57.
2. Anderson, J.S., and M.I. Ritchie: Solution Mining of Uranium. Min. Cong. J., V. 54, No. 1, January 1968, pp. 20–26.
3. Anderson, M.L.: Improved Productivity from In Situ Leaching. Proc. on Productivity

in Mining, Univ. Missouri—Rolla, Rolla, Mo., May 13–15, 1974, Ed. by J.J. Scott, pp. 82–83.
4. Arnold, W.D., and D.J. Crouse: Radioactive Contamination of Copper Recovered from Ore Fractured with Nuclear Explosions. Oak Ridge National Laboratory, Rept. ORNL-4677, September 1971, 55 pp.
5. Austin, W.L.: Process of Extracting Copper from Ore. U.S. Pat. 975,106, Nov. 8, 1910, 2 pp.
6. Ballard, J.K.: Solution Mining. Mining Engineering, V. 23, No. 2, February 1971, p. 109.
7. Becherkin, S.G., V.G. Bakhurov, and I.K. Lutsenko.: Underground Leaching of Uranium from Low-Grade In Situ Ores. Trans. Atomic Energy, V. 21, No. 2, February 1968, pp. 128–133.
8. Braun, R.L., A.E. Lewis, and M.E. Wadsworth: In-Place Leaching of Primary Sulfide Ores: Laboratory Leaching Data and Kinetics Model. Proc. Solution Mining Symp., AIME Annual Meeting, Dallas, Texas, Feb. 25–27, 1974, pp. 295–323.
9. Braun, R.L., R.G. Mallon, and A.E. Lewis: Nuclear Solution Mining: Oxygen Distribution. Lawrence Livermore Laboratory. To be published.
10. Brennan, C.V.: Mining Operations at the Property of the Britannia Mining and Smelting Company. BUMINES RI 6815, January 1935, 36 pp.
11. Bruynesteyn, A., and D.W. Duncan: Effect of Particle Size on the Microbiological Leaching of Chalcopyrite Bearing Ore. Proc. Solution Mining Symp., AIME Annual Meeting, Dallas, Texas, Feb. 25–27, 1974, pp. 324–337.
12. Carnahan, T.G., and H.J. Heinen: Simulated In Situ Leaching of Copper from a Porphyry Ore. BUMES TPR 69, May 1973, 11 pp.
13. Crawford, E.: Developers Eye Texas Potential for In Situ Uranium Leaching. Eng. and Min. J., V. 176, No. 7, July 1975, pp. 81–82. TWIN CITIES MINING RESEARCH CENTER P.O. BOX 1660 TWIN CITIES, MINNESOTA 55111.
14. Dandrea, D.V.: Zonia Mine, Notes from Aug. 16, 1973 Visit to Zonia Mine, Arizona, 1973, 3 pp.; Available from Consultation at Bureau of Mines, Twin Cities Mining Research Center, Minneapolis, Minnesota.
15. Dandrea, D.V., R.A. Dick, R.C. Steckley, and W.C. Larson: A Fragmentation Experiment for In Situ Extraction. Proc. Solution Mining Symp., AIME Annual Meeting, Dallas, Texas, Feb. 25–27, 1974, pp. 148–161.
16. Dandrea, D.V., and W.C. Larson: Phase I In Situ Leach Test at the Emerald Isle Mine. BUMINES Internal Report, 1975, 77 pp.; Available for Consultation at Bureau of Mines, Twin Cities Mining Research Center, Minneapolis, Minnesota.
17. Dandrea, D.V., and W.C. Larson.: Phase II In Situ Leach Test at the Emerald Isle Mine—BUMINES Internal Report, 1975, 24 pp.; Available for Consultation at Bureau of Mines, Twin Cities Mining Research Center, Minneapolis, Minnesota.
18. Dew, J.N., and W.L. Martin (Assigned to Continental Oil Company): In Situ Roasting and Leaching of Uranium Ores. U.S. Pat. 2,954,218, Sept. 27, 1960, 3 pp.
19. Dick, R.A.: In Situ Fragmentation for Solution Mining—A Research Need. Pres. at the Second Internat. Symp. on Drilling and Blasting, Phoenix, Arizona, Feb. 12–16, 1973.; Available for Consultation at Bureau of Mines, Twin Cities Mining Research Center, Minneapolis, Minnesota.
20. Dick, R.A., et al.: Porphyry Copper—In Situ Recovery by Leaching Following Non-Nuclear Fracturing. Interdisciplinary Research Task Force, November 1971, 20 pp.; Available for Consultation at Bureau of Mines, Twin Cities Mining Research Center, Minneapolis, Minnesota.
21. Dravo Corporation: Solution Mining. Section 6.6.2 In Analysis of Large Scale Non-Coal Underground Mining Methods, USBM Contract Report S0122059, January 1974, pp. 455–464.
22. Dudas, L., H. Mass, and R. Bhappu: Role of Mineralogy in Heap and In Situ Leaching of Copper Ores. Proc. Solution Mining Symp., AIME Annual Meeting, Dallas, Texas, Feb. 25–27, 1974, pp. 193–209.
23. Duncan, D.W., and A. Bruynesteyn: Enhancing Bacterial Activity in a Uranium Mine Can. Min, and Met. Bull., May 1971, pp. 32–36.
24. Duncan, D.W., and A. Bryunesteyn: Microbiological Leaching of Uranium. Proc. 1970 Uranium Symp., Socorro, New Mexico, Comp. by R.J. Roman, pp. 55–61.

25. Duncan, D.W., C.C. Walden, P.C. Trussell, and E.A. Lowe: Recent Advances in the Microbiological Leaching of Sulfides. Trans. Soc. Min. Eng., V. 238, June 1967, pp. 122–128.
26. Engineering and Mining Journal: AEC and KCC Will Jointly Study Potential of Nuclear Blasting to Mine Copper Ore. V. 174, No. 4, April 1973, pp. 26–30.
27. Engineering and Mining Journal: ASARCO and DOW Chemical to Leach Deep Copper Orebody In Situ in Arizona. V. 173, No. 6, June 1972, p. 19.
28. Engineering and Mining Journal: Copper Leaching with Cyanide—A Review of Five Inventions. V. 168, No. 9, September 1967, pp. 123–127.
29. Engineering and Mining Journal: The Cost of Leaching U_3O_8 in Stopes and Pumping Solution to the Surface. V. 169, No. 6, June 1968, p. 186.
30. Engineering and Mining Journal: The Estimated Cost of a Nuclear Leaching Experiment on Copper Ore. V. 169, No. 6, June 1968, p. 186.
31. Engineering and Mining Journal: India Studies Possibility of Copper Mining by Nuclear Blasts. V. 171, No. 7, July 1970, 33 pp.
32. Engineering and Mining Journal: Kennecott Investigates Solution Mining Potential of Deep Cu Deposit. V. 176, No. 9, September 1975, pp. 37, 41.
33. Engineering and Mining Journal: Kennecott Sets Sights on Nuclear Test for In Situ Recovery of Copper. V. 168, No. 11, November 1967, pp. 116–122.
34. Engineering and Mining Journal: Mining In Situ by Nature's Easy Way. V. 168, No. 10, October 1967, pp. 75–80.
35. Engineering and Mining Journal: New Method Suggested for Leaching Uranium Ore in Place. V. 168, No. 5, May 1967, pp. 106–107.
36. Engineering and Mining Journal: Ranchers Big Blast Shatters Copper Orebody for In Situ Leaching. V. 173, No. 4, April 1972, pp. 98–100.
37. Engineering and Mining Journal: Solution Mining Opening New Reserves. V. 175, No. 7, July 1974, pp. 62–71.
38. Engineering and Mining Journal: Texas-Uranium Production by In Situ Leaching. V. 176, No. 6, June 1975, p. 304.
39. Finlay, W.L.: Molecular Mining of Keweenaw Copper, A Prime Candidate for the Enlargement of Non-Renewable Mineral Reserves through Technology. Panel Discussion at the Joint NAS-NAE Meeting on National Policy, Washington, D.C., Oct. 25, 1973, 15 pp.
40. Fisher, J.R.: Bacterial Leaching of Elliot Lake Uranium Ore. Can. Min. and Met. Bull., May 1966, pp. 588–592.
41. Fisher, J.R., F.C. Lendrum, and B.G. Macdermid: Laboratory and Underground Studies of Bacterial Leaching of Elliot Lake Ores. AIME Annual Meeting, 1966, 9 pp.
42. Fitch, J.L., and B.G. Hurd (Assigned to Mobil Oil Corporation): In Situ Leaching Method. U.S. Pat. 3,278,232, Oct. 11, 1966, 5pp.
43. Fitch, J.L., and B.G. Hurd (Assigned to Mobil Oil Corporation): Method of Leaching Sub-Surface Minerals In Situ. U.S. Pat. 3,309,141, Mar. 14, 1967, 8 pp.
44. Fletcher, J.B.: In Place Leaching. Skillings Mining Review, V. 63, No. 17, Apr. 27. 1974, pp. 7–10.
45. Fletcher, J.B.: In Place Leaching, Miami Mine. Pres. at Arizona Section Meeting, Milling Div., AIME, Apr. 6, 1962, 9 pp.
46. Fletcher, J.B.: In Place Leaching—Miami Mine, Miami, Arizona. Pres. at AIME Centennial Annual Meeting, Soc. of Min. Eng., AIME, New York, N.Y., March 1971, AIME Preprint 71-AS-40; Trans. Soc. of Min. Eng., AIME, V. 250, No. 4, December 1971, pp. 310–314.
47. Gardner, J., and M.I. Ritchie (Assigned to Utah Construction and Mining Company): Leaching of Uranium Ore In Situ. U.S. Pat. 3,309,140, Mar. 14, 1967, 8 pp.
48. George, D.R., and J.R. Ross. Recovery of Uranium from Uranium Mine Waters and Copper Ore Leaching Solutions, Processing of Low-Grade Uranium Ores. Proc. of a Panel Held in Vienna, June 27–July 1, 1966. Internat. Atomic Energy Agency, Vienna, 1967, pp. 227–234.
49. Greenwood, C.C.: Underground Leaching at Cananea. Eng. and Min. J., V. 121, No. 13, March 1926, pp. 518–521.
50. Grimes, M.E.: Prediction of Uranium Extraction in In Situ Stope Leaching. Proc.

Solution Mining Symp., AIME Annual Meeting, Dallas, Texas, Feb. 25–27, 1974, pp. 338–353.
51. Griswold, G.B.: Rock Fracturing Techniques for In Place Leaching. Pres. Annual Meeting of the Soc. of Min. Eng., AIME, Washington, D.C., Feb. 16–20, 1969, AIME Preprint 69-AS-76, 4 pp.
52. Groves, R.D., T.H. Jeffers, and G.M. Potter: Leaching Coarse Native Copper Ore with Dilute Ammonium Carbonate Solution, Proc. Solution Mining Symp., AIME Annual Meeting, Dallas, Texas, Feb. 25–27, 1974, pp. 381–389.
53. Gubkin, N.V., D.T. Desyatnikov, and I.K. Rudneva: Advantages of Mining Uranium in Inundated Sheet Deposits by Underground Leaching. Atomic Energy, V. 24, No. 6, June 1968, pp. 511–514.
54. Hansen, S.M.: Nuclear Blasting for Mining and Leaching. World Mining, V. 1, No. 1, September 1965, pp. 20–27.
55. Hansen, S.M., and A.R. Jager: How to Make Ore from Marginal Deposits. Eng. and Min. J., V. 169, No. 12, December 1968, pp. 75–81.
56. Hard, R.A., and R.L. Ripley (Assigned to Union Carbide Corporation): Process for Leaching Base Elements, Such as Uranium Ore In Situ. U.S. Pat. 3,708,209, Jan. 2, 1973, 10 pp.
57. Hardwick, W.R.: Fracturing a Deposit with Nuclear Explosives and Recovering Copper by the In Situ Leaching Method. BUMINES RI 6996, 1967, 48 pp.
58. Harrison, V.F., W.A. Gow, and M.R. Hughson: Part I. Factors Influencing the Application of Bacterial Leaching to a Canadian Uranium Ore. J. of Metals, V. 18, No. 11, November 1966, pp. 1189–1194.
59. Harrison, V.F., W.A. Gow, and M.R. Hughson: Leaching of Uranium From Elliot Lake Ore in the Presence of Bacteria. Can. Min. J., V. 87, No. 5, May 1966, pp. 64–67.
60. Heinen, H.J., T.G. Carnahan, and J.A. Eisele (Assigned to U.S. Department of the Interior): Chemical Mining of Copper Porphyry Ores. U.S. Pat. 3,890,007, June 17, 1975, 6 pp.
61. Heinen, H., and B. Porter: Experimental Leaching of Gold from Mine Waste. BUMINES RI 7250, April 1969, 5 pp.
62. Hogan, D.K.: In Situ Copper Leaching at the Old Reliable Mine. Mining Magazine, V. 24, No. 5, May 1974, pp. 353–359.
63. Holderreed, F.L.: Copper Extraction by the Acid Leaching of Broken Ore. Report of Pincock, Allen and Holt, Inc., 1975, 19 pp.
64. Hunkin, G.G.: A Review of In Situ Leaching. Pres. at AIME Annual Meeting, Soc. of Min. Eng., AIME, New York, N.Y., Feb. 26–Mar. 4, 1971, AIME Preprint 71-AS-88, 23 pp.
65. Hurd, B.G., and J.L. Fitch (Assigned to Mobil Oil Corporation): In Situ Leaching of Subterranean Deposits. U.S. Pat. 3,278,233, Mar. 14, 1967, 10 pp.
66. Johnson, P.H.: Acid-Ferric Sulfate Solutions for Chemical Mining. Mining Engineering, V. 17, No. 8, August 1965, pp. 64–68.
67. Johnson, P.H., and R.B. Bhappu: Chemical Mining—A Study of Leaching Agents. New Mexico Bureau of Mines and Mineral Resources, Circ. 99, 1969, 7 pp.
68. Johnson, P.H., and R.B. Bhappu: Chemical Mining—Theoretical and Practical Aspects. Pres. at SME Fall Meeting—Rocky Mountain Minerals Conf., Las Vegas, Nevada, Sep. 6–8, 1967, 20 pp.
69. Kaczynski, D.A., G.W. Lower, and W.A. Hockings: Kinetics of Leaching Metallic Copper in Aqueous Cupric Ammonium Nitrate Solutions. Proc. Solution Mining Symp., AIME Annual Meeting, Dallas, Texas, Feb. 25–27, 1974, pp. 390–400.
70. Kalabin, A.I.: Winning of Useful Elements from Minerals by Leaching Underground. Mining Magazine, V. 118, February 1968, pp. 129–134.
71. Keyes, H.E.: Discussion of In Place Leaching at Miami Mine, Miami, Arizona, by J.B. Fletcher, in Trans. Soc. of Min. Eng., V. 250, No. 4, December 1971, pp. 310–314. Trans. Soc. of Min. Eng., V. 252, No. 2, June 1972, pp. 186–187.
72. Lackey, J.A.: Solution Mining (In Situ Leaching) A Literature Survey. Amdel Bull. (The Australian Mineral Development Laboratories), No. 19, April 1975, pp. 40–61.
73. Lampard, W.J. (Assigned to Kennecott Copper Corporation): Controlled In Situ

Leaching of Ore Deposits Utilizing Pre-Split Blasting. U.S. Pat. 3,863,987, Feb. 4, 1975, 6 pp.
74. Lang, E.A., and W.R. Archibald: Innovative Systems for the Recovery of Uranium. Pres. at the National Western Min. Conf., Denver, Colo., Feb. 7, 1975, pp. 1–13; Colorado Mining Assoc., 1975 Mining Yearbook, pp. 114–118.
75. Larson, D.R.: A Report Comparing Open Pit Mining and Heap Leaching with an In Situ Leaching System for a Hypothetical Oxide Copper Deposit Based on the Characteristics of the Cactus Deposit, Miami, Arizona. M.S. Thesis, Univ. Minnesota, Dept. of Mineral and Metallurgical Eng., June 1973, 254 pp.
76. Laswell, G.W.: Considerations Applicable to Utilization of Rotary Drilling for In Place Recovery of Minerals and Hydrocarbons. Report for Pincock, Allen and Holt, Inc., Feb. 1975, 7 pp.
77. Lewis, A.E.: Chemical Mining of Primary Copper Ores by Use of Nuclear Technology. Proc. Symp. of Engineering with Nuclear Explosives, Las Vegas, Nev., Jan. 14–16, 1970, Conf. 700101, V. 21, pp. 907–917.
78. Lewis, A.E., and R.L. Braun—Nuclear Chemical Mining of Primary Copper Sulfides. Pres. at the AIME Meeting, San Francisco, Calif., Jan. 27, 1972, 30 pp.
79. Lewis, A.E., R.L. Braun, C. J. Sisemore, and R.G. Mallon: Nuclear Solution Mining-Breaking and Leaching Considerations. Proc. Solution Mining Symp., AIME Annual Meeting, Dallas, Texas, Feb. 25–27, 1974, pp. 56–75.
80. Lewis, F.M., and R.B. Bhappu: Evaluating Mining Ventures via Feasibility Studies. Mining Engineering, V. 27, No. 10, October 1975, pp. 48–54.
81. Livingston, C.W.: Method of Mining Ores In Situ by Leaching. U.S. Pat. 2,818,240, Dec. 31, 1957, 14 pp.
82. Longwell, R.L.: In Place Leaching of a Mixed Copper Ore Body. Proc. Solution Mining Symp., AIME Annual Meeting, Dallas, Texas, Feb. 25–27, 1974, pp. 233–242.
83. MacGregor, R.A.: Recovery of U_3O_8 by Underground Leaching. Pres. at Annual General Meeting, Montreal, April 1964, Can. Min. and Met. Bull., May 1966, pp. 583–587.
84. MacGregor, R.A.: Uranium Dividends from Bacterial Leaching. Mining Engineering, V. 21, No. 3, March 1969, pp. 54–55.
85. Malouf, E.E.: Copper Leaching Practices. Soc. of Min. Eng., AIME, AIME Preprint 72-AS-84, 1972, 7 pp.
86. Malouf, E.E.: Introduction to Dump Leaching Practice. Part I of Second Tutorial Symp. on Extractive Metallurgy, Univ. Utah, Dec. 14–16, 1972, 26 pp.
87. Malouf, E.E.: The Role of Microorganisms in Chemical Mining, Mining Engineering, V. 23, No. 11, November 1971, pp. 43–46.
88. Marrs, L.F.: Underground Leaching of Uranium at the Pitch Mine. Pres. American Min. Cong. Convention, Denver, Colo., Sept. 27–30, 1970, 14 pp.; Min. Cong. J., V. 56, No. 11, November 1970, pp. 35–43.
89. Martin, J.W. (Assigned to Oil Recovery Corporation): Recovery of Uranium by Carbonated Water and Surface Action Agents and Wetting Agents. U.S. Pat. 3,130,960, Apr. 28, 1964, 8 pp.
90. McKinney, W.A.: Solution Mining. Mining Engineering, V. 25, No. 2, February 1973, pp. 56–57.
91. Michigan Technological University: A Plan for Determining the Feasibility of an In Situ Leaching of Native Copper Ores. Prepared for Homestake Copper Company, Calumet, Michigan, June 14, 1974, 35 pp.
92. Miller, J.D.: Processing of Leach Liquors Produced by Nuclear Solution Mining. Lawrence Livermore Laboratory, Rept. UCRL-51350, Feb. 15, 1973, 18 pp.
93. Mining Congress Journal: An Experimental In Place Leaching Test. V. 59, No. 5, May 1973, p. 10.
94. Mining Engineering. Kennecott Proposes Nuclear Mining Experiments at Safford Deposit. V. 19, No. 11, November 1967, pp. 66–67.
95. Mining Engineering: Ranchers Development Sets off Blast, Will Leach at Big Mike. V. 25, No. 8, August 1973, p. 10.
96. Mining Magazine: In Situ Leaching—A Review. V. 125, No. 3, September 1971, pp. 213–215.

97. Mining Record (Denver, Colorado): Ranchers Blasts Old Reliable to Test New Mining Technique. V. 83, No. 12, Mar. 22, 1972, p. 1.
98. Myers, D.L.: Mining Copper In Situ. Mines Mag., V. 31, No. 6, June 1941, pp. 255–263.
99. Pinge, W.B.: Bacterial Leaching. Mineral Industries Bull., V. 2, No. 3, May 1968, pp. 1–19.
100. Pojar, M.: In Place Leaching of a Copper Sulfide Deposit. Pres. at Duluth Meeting, Soc. of Min. Eng., AIME, Duluth, Minn., Jan. 16–18, 1974, 9 pp.
101. Popoff, D.R.: Bacterial Leaching Means New Life for Mine. Canadian Nuclear Technology, V. 5, No. 4, July–August 1966, pp. 31–33.
102. Porter, D.D.: Blast Design for In Situ Leaching. Pres. at Southwest Mineral Industry Conf., Phoenix, Ariz., Apr. 27, 1973, 7 pp.; Available from E.I. Du Pont de Nemours and Co.; Inc.; Wilmington, Del. 19898.
103. Porter, D.D., and H.G. Carlevato. In Situ Leaching: A New Blasting Challenge. Proc. Solution Mining Symp., AIME Annual Meeting, Dallas, Texas, Feb. 25–27, 1974, pp. 33–43.
104. Potter, G.M.: Solution Mining. Mining Engineering, V. 21, No. 1, February 1969, pp. 68–69.
105. Prescott Newspaper (Prescott, Arizona): Kirkland Blast will Open new Body of Copper. Apr. 18, 1973, p. 1.
106. Rabb, D.D.: Leaching of Copper Ores and the Use of Bacteria. Lawrence Radiation Laboratory, Rept. UCID-4958, Jan. 22, 1965, 25 pp.
107. Rabb, D.D.: Penetration of Leach Solution into Rocks Fractured by a Nuclear Explosion. Annual Meeting of Soc. of Min. Eng., AIME, Washington, D.C., Feb. 16–20, 1967 V. 1:A, AIME Preprint 69-AS-4, 14 pp.
108. Rabb, D.D.: Solution Mining. Mining Engineering, V. 24, No. 2, February 1972, pp. 62–64.
109. Ranchers Exploration and Development Corp.: The Old Reliable Project—A Test of New Technology. Company Report, 4 pp.
110. Rickard, R.S.: Chemical Constraints on In Situ Leaching and Metal Recovery. Proc. Solution Mining Symp., AIME Annual Meeting, Dallas, Texas, Feb. 25–27, 1974, pp. 185–192.
111. Roman, R.J.: Solution Mining. Mining Engineering, V. 26, No. 2, February 1974, pp. 43–44.
112. Rosenbaum, J.B., and W. A. McKinney: In Situ Recovery of Copper from Sulfide Orebodies Following Nuclear Fracturing. Proc. Symp. of Engineering with Nuclear Explosives, Las Vegas, Nev., Jan. 14–16, 1970, Conf. 700101, V. 2, pp. 877–887.
113. Rosenzweig, M.D.: New Copper Technology is Winning the Ore. Chemical Eng., Dec. 4, 1967, pp. 88–90, 92.
114. Rouse, J.V.: Environmental Aspects of In Situ Mining and Dump Leaching. Proc. Solution Mining Symp., AIME Annual Meeting, Dallas, Texas, Feb. 25–27, 1974, pp. 3–14.
115. Rudershausen, C.G.: Copper Solution Mining at Old Reliable. Pres. at National Meeting of American Inst. of Chemical Eng., Salt Lake City, Utah., Aug. 18–21, 1974, 11 pp.
116. St. Peter, A.L.: In Situ Leaching of Orebodies Design and Management. Report for Pincock, Allen and Holt, Inc., 1975, 20 pp.
117. Scott, W.G. (Assigned to Inspiration Consolidated Copper Company): Leaching Copper Ores. U.S. Pat. 2,563,623, Aug. 7, 1951, 14 pp.
118. Sheffer, H.W., and L.G. Evans: Copper Leaching Practices in the Western United States. BUMINES IC 8341, 1968, 57 pp.
119. Shock, D.A.: Developments in Solution Mining Portend Greater Use for In Situ Leaching. Mining Engineering, V. 22, No. 2, February 1970, p. 73.
120. Shock, D.A., and F.R. Conley: Solution Mining—Its Promise and its Problems. Proc. Solution Mining Symp., AIME Annual Meeting, Dallas, Texas, Feb. 25–27, 1974, pp. 79–97.
121. Shock, D.A., and J.G. Davis: Hydrofracing as a Mining Technique. Soc. of Min. Eng., AIME, 1969, AIME Preprint 69-AS-63, 25 pp.
122. Shoemaker, R.S.: Ammonia Revival for the Keweenaw. Mining Enginnering, V. 24, No. 5, May 1972, pp. 45–47.

123. Sievert, J.A., W.L. Martin, and F.R. Conley. In Situ Leaching of Uranium. Pres. at Soc. Min. Eng. Fall Meeting, St. Louis, Mo., 1970, 18 pp.
124. Skillings Mining Review: El Paso Studying In Situ Leaching at Emerald Isle. V. 63, No. 52, Dec. 28, 1974, p. 8.
125. Skillings Mining Review: Ranchers Detonates Large Blast at Big Mike Mine in Nevada. V. 62, No. 29, July 21, 1973, p. 5.
126. Spedden, H.R., E.E. Malouf, and J. Davis: In Situ Leaching of Copper—Pilot Plant Test. Pres. at Ann. Meeting, AIME, New York, February 1971, 21 pp.
127. Steckley, R.C., W.C. Larson, D.V. Dandrea: Blasting Tests in a Porphyry Copper Deposit—In Preparation for In Situ Extraction. BUMINES RI 8070, 1975, 57 pp.
128. Stewart, R.M., and K.E. Niermever: Nuclear Technology and Mineral Recovery. Proc. Symp. on Engineering with Nuclear Explosives, Las Vegas, Nev., Jan. 14–16, 1970, Conf. 700101, V. 2, pp. 864–876.
129. Tewes, H.A., H.B. Levy, and L.L. Schwartz—Nuclear Chemical Copper Mining and Refining: Radiological Considerations. Lawrence Livermore Laboratory, Rept. UCRL-51345, Rev. 1, June 3, 1974, 34 pp.
130. Thomas, R.W.: Leaching Copper from Worked-Out Areas of the Ray Mines, Arizona Min. and Met., V. 19, November 1938, pp. 481–485.
131. Trussell, P.C., D.W. Duncan, and C.C. Walden: Biological Mining. Can. Min. J., V. 85, No. 3, March 1964, pp. 46–49.
132. Tuovinen, O.H.: Microbiological Aspects in the Leaching of Uranium by Thiobacillus Ferrooxidans. Atomic Energy Review, V. 10, No. 2, 1972, pp. 251–258.
133. U.S. Atomic Energy Commission: In Situ Uranium Leaching Bibliography. November 1974, 8 pp.
134. Ward, M.H.: Engineering for In Situ Leaching. Min. Cong. J., V. 59, No. 1, January 1973, pp. 21–27.
135. Ward, M.H.: Surface Blasting Followed by In Situ Leaching the Big Mike Mine. Proc. Solution Mining Symp., AIME Annual Meeting, Dallas, Texas, Feb. 25–27, 1974, pp. 243–251.
136. Weed, R.C.: Cananeas Program for Leaching In Place. Mining Engineering, V. 8, No. 7, July 1, 1956, pp. 721–723.
137. White, L. (Ed.): In Situ Leaching Opens New Uranium Reserves in Texas. Eng. and Min. J., V. 176, No. 7, July 1975, pp. 73–80.
138. Wormser, F.E.: Leaching a Copper Mine. Eng. and Min. J., V. 116, No. 16, October 1923, pp. 665–670.
139. Zimmer, P.F., and M.A. Lekas (Comp. and Ed. by). Project Sloop (Nuclear Explosives-Peaceful Applications): U.S. Atomic Energy Commission Report No. PNE 1300, June 1, 1967, 44 pp.

Paper 6

RECENT DEVELOPMENTS IN MINERAL PROCESSING AND THEIR ECONOMIC IMPLICATIONS

by D. N. Moir

THE NEED TO DEVELOP the correct and most economic flowsheet for the beneficiation of mineral ores is greater now than ever before due to the rapid rise in costs of labour, power and capital equipment, and the generally declining ore grades and hence higher rates of treatment which have to be employed. In addition, finer-grained more complex ores are being treated and consumers are demanding higher product purities in order that their costs are minimised. In considering the various unit processes the trend towards larger and larger units continues, this is particularly true in the case of grinding mills and flotation cells. This is done in order to minimise capital and running costs by reducing the total number of motors, pumps and spare parts which have to be provided and to reduce the total space occupied, quite apart from the fact that the larger units may be more efficient in themselves. A problem is that larger single unit stages are less flexible and represent large capital expenditure on few items; it is therefore essential to ensure that the expected efficiency will be attained in practice. An example is given later of the scale of testwork required to decide what type of grinding circuit would prove to be the most economic for a large iron ore beneficiation operation.

Screening Systems

Basically the requirement of any screening system is to be able to produce the required size fraction or fractions at the highest possible rate and at the same time produce as sharp a separation as possible for the lowest cost. The developments are therefore directed towards achieving these objectives. There are two ways in which these objectives can be realised: (1) by making improvements to conventional screening systems in which the material is presented to the screen in the normal way, or (2) by adopting the so called "probability" screening principle in which use is made of the fact that a stream of particles moving rapidly, almost at right angles to a screening surface, will only pass through the screen when the particle size is about ½d, where d is the distance between the screening elements. This is typified by the Dutch State Mines (DSM) screen or sieve bend, in which a screen surface composed of wedge wires has been curved into a quarter circle, the length of the wires being at right-angles to the flow of material in the form of a slurry. The DSM sieve bend has many applications, including the removal of heavy media from products, closing coarse-size grinding circuits and screening final products.

However, the system suffers some disadvantages in that the wedge wires tend to become blocked, particularly if there is a large proportion of angular particles in the feed which are similar in size to that of the screen spacing. Wear of the profile of the wedge wires can also affect performance. In order to overcome this tendency, devices for rapping or vibrating the screen have been added, one such device being the "Rapifine" screen. The rapping action helps to remove the particles caught in the wedge wires and to detach the "through screen" material which tends to cling to the underside of the screen.

This problem of blinding and detachment of through-screen material has been dealt with in the design of the Bartles-CTS screen,[1] which consists of a short-radius curved screen cloth fitted with "crimps" on its under surface, see Fig. 1. The screen is usually constructed in two stages as shown, feed being applied tangentially in a thin laminar film to the upper section through which approximately half the undersize passes. The second section then re-treats the oversize from the first section, removing the remainder of the undersize. The same relationship of passing size to screen aperture applies as with the DSM screen.

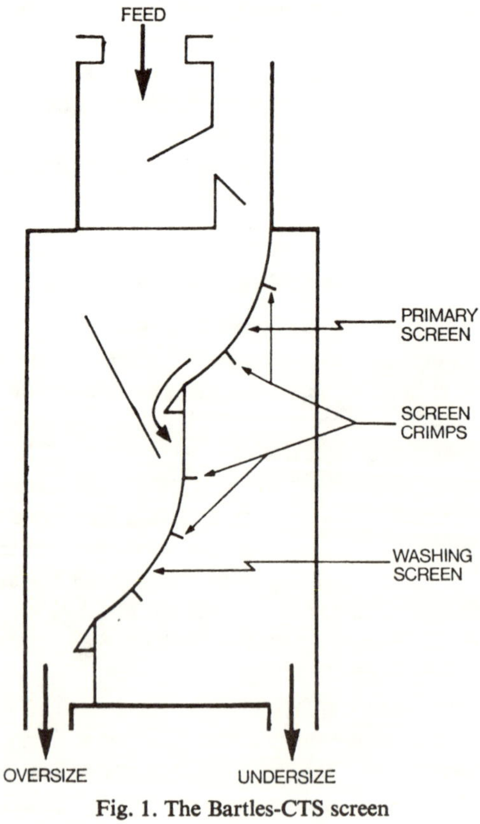

Fig. 1. The Bartles-CTS screen

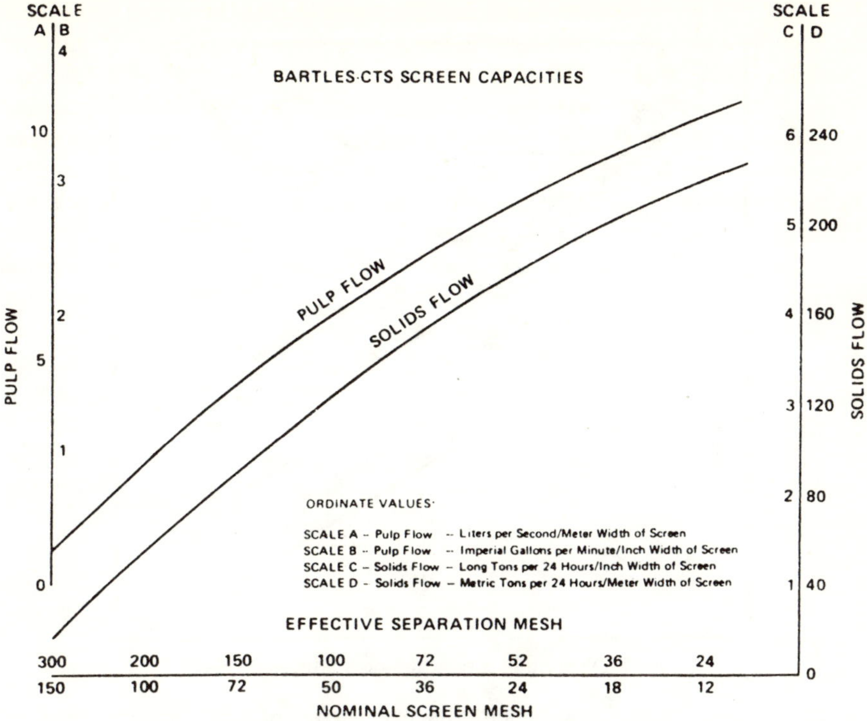

Fig. 2. Capacity vs separation mesh for Bartles-CTS screens

An alternative configuration is a quadruplex system in which the upper pair of screens have a coarser aperture than the lower pair, so that coarse material is removed first, thus reducing the duty of the lower, finer screens which make the final separation. It is possible to obtain three products from such a system.

Data given by the manufacturers suggest that screening a 20% solids pulp consisting of particles of specific gravity 2.7, with a natural size distribution, a throughput rate of 5 tons per 24 hours per inch width of screen (200 tonnes/24h/m) can be achieved cutting at 36 mesh,* and 1 ton per 24 hours per inch width (40 tonnes/24h/m) at 200 mesh. Additional data may be read from Fig. 2. The screens are made in standard widths of 16, 34 and 68 inches (40, 86 and 173 cm). The cost of replacing screen cloth is said to be of the order of 0.5 to 0.7 pence per tonne. The cost of a cloth for a duplex screen is approximately £10 and in some cases a throughput of over 4000 tonnes has been achieved before replacement was needed. Blinding is also said to be less than with sieve bends, full data is given in Ref. 1.

The screens are currently being used in the removal of primary fines in crushing plants, in dense media separations, closed circuit grinding, rejection

* BSS meshes are used throughout this paper.

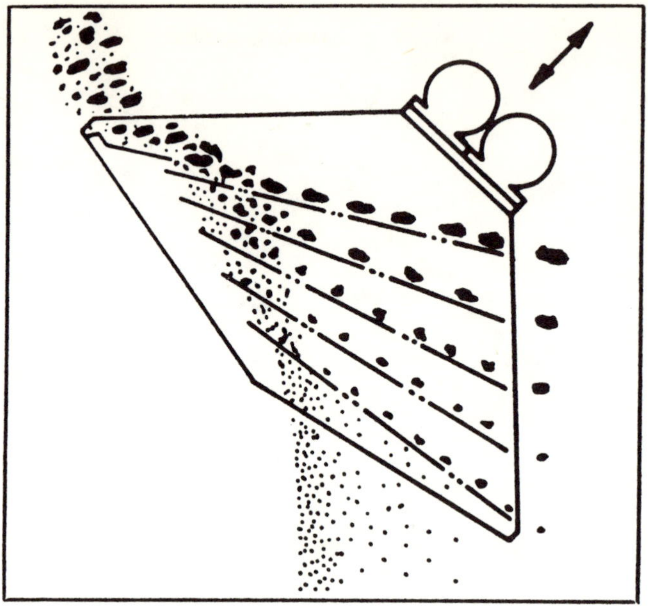

Fig. 3. Diagram of a typical 5-deck Mogensen sizer

of tramp oversize in flotation circuits, wet sizing of final products and in numerous other applications. Capital cost of a 68 inch (1.727 m) screen was approximately £725 ex works in October 1975. The specific capital cost is thus £25/tonne/h screening at 30 mesh or £150/tonne/h when screening at 200 mesh.

Another method of screening using the probability principle is the Mogensen sizer[2] which consists of a number of screen decks, commonly fine, set in the form of a fan as shown in Fig. 3. The meshes are finer at the bottom and coarser at the top (feed) and the meshes are chosen so that ideally the particles arriving at any given deck are smaller than the apertures in that deck. As with the sieve bend, the separations made by the screen decks are smaller than the apertures in the deck. Material is fed onto the screens in such a way as to ensure that the particles move individually and freely within the machine, the principal method of travel is vertically through the machine and not horizontally across the screens. This type of screen generally uses less power than its conventional counterparts, is lighter and less susceptible to blinding. Data made available by the company show that treating hot sand sized 4.5 mm to 0 mm at 90–100°C and screening to remove +2 mm particles, the feed rate is 8–10 tonnes/h for a power consumption of 2 kW. Screens were changed 10 times for 87,000 tonnes throughput at a total cost of £170 giving a cost of 0.19 pence/tonne (1971). Life of the screens was about 30 times longer than that of conventional screens. Screening aluminium oxide grains having

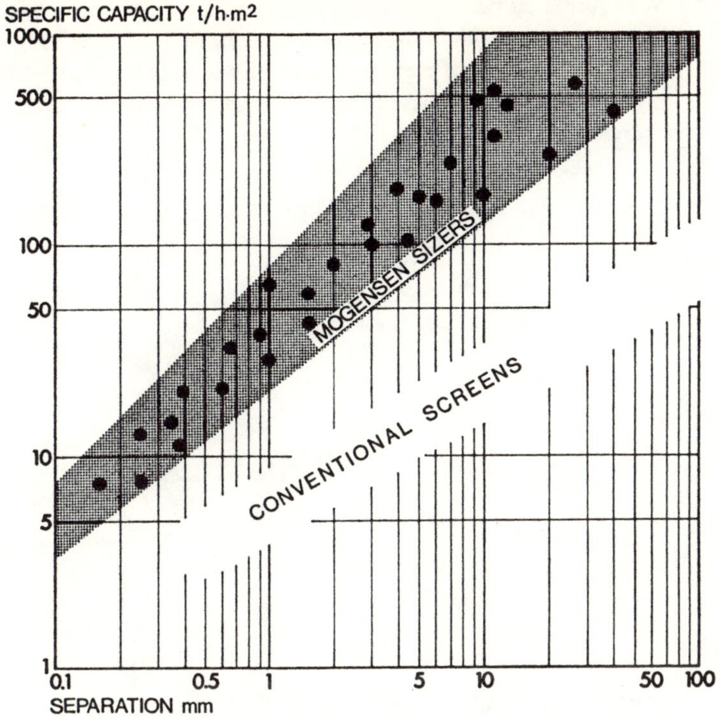

Fig. 4. Capacity vs separation size for Mogensen sizers

a moisture content of 8–10%, sized 1.0 mm to 0 mm and cutting at 85 mesh (0.18 mm), 4 to 5 tonnes/h were achieved for a power consumption of 2 kW. Screen replacement cost was 2.0 pence/tonne.

This type of screening machine is versatile, it can handle dry, wet or moist materials, can screen in the range 2 inches to 150 mesh (50–0.1 mm), can deal with up to 350 tonnes/h, can deal with hot materials up to 800°C and can be made totally enclosed for dry fine-sizing applications. A rough guide to the throughput of the sizer can be obtained from Fig. 4 where its performance is compared with conventional screening.

In the HCC-Burstlein screening system[3] a different approach has been taken. The thickness of the screen bed has been kept constant or even increased slightly from feed to discharge end of the screen system by adjusting the slope of the screen units. In the normal screening system the particle bed depth generally decreases from the feed end to the discharge end as the fines are removed. Thus at the feed end the bed is too thick to allow the finer particles, present in large numbers, to penetrate the bed and reach the screening surface. Towards the discharge end where the bed thins out there are not sufficient numbers of fine particles being presented to the screen surface and its potential capacity is thus not utilised. For efficient screening therefore the feed end bed requires to be thinned out and the discharge end thickened in order to achieve the best effect. Burstlein has done this by breaking up the screen surface into a number of elements each of which is inclined at the correct angle and vibrated at the correct frequency and direction to achieve this effect the arrangement is shown in Fig. 5.

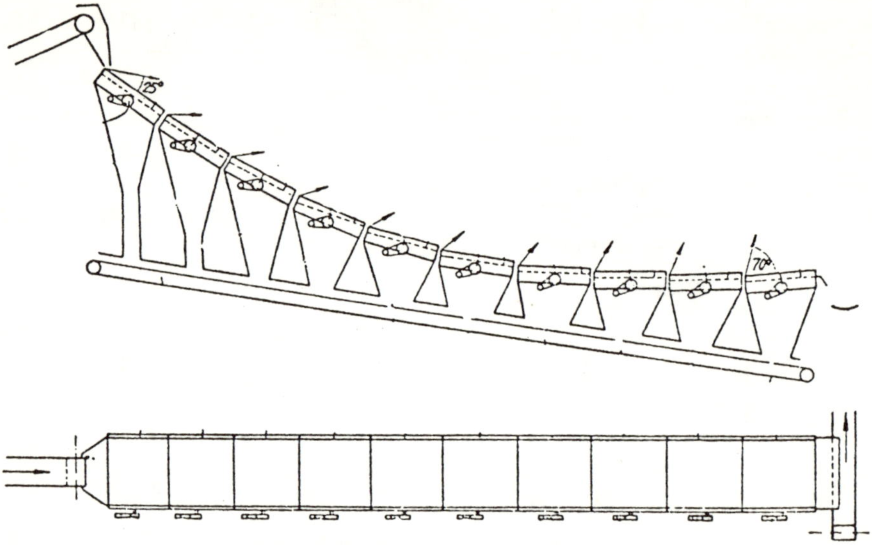

Fig. 5. HCC Burstlein screening system

It is claimed that the specific screening rate is between two and three times that for conventional screens, thus for a given screening rate, a reduction of the total screening surface can be made. An additional advantage is that the screening area of an HCC installation can be made as large as 100 m^2 (1076 ft^2) compared with the normal 20 m^2 (215 ft^2) of conventional screen installations, thus avoiding the problems of multiple stream division and consequent high installation costs.

All in all the capital, operating and maintenance costs of an HCC installation are about 50% of those of conventional screening installations.

As an example, in an iron ore screening application, 5000 tonnes/h having a size range of 250 mm to 0 mm were screened at 5 mm (0.19 inches) using 4 HCC screens of 60 m^2 (645 ft^2) each. Conventional screening required 42 screens of 15 m^2 (161 ft^2) each. Thus, making a reduction in screening area from a total of 620 m^2 (6671 ft^2) to 240 m^2 (2582 ft^2), and by reducing the number of screening units from 42 to 4 with consequent reduction in the number of process streams, a saving of 50% of the capital, operating and maintenance costs was made. So far the chief applications for this process seem to have been in screening iron ore and charges for blast furnaces.

Comminution

In most concentration plants involving fine grinding, the largest power consumption is almost certain to be in the grinding section where operating costs are frequently from 30% to as much as 50% of the total cost of processing. It is not surprising therefore that considerable efforts have been made to develop different milling methods and machinery, in order to reduce power and capital costs and the associated running costs—which include those of grinding media, liners, maintenance, labour etc—and to produce a better sized product, ie one with a closer size distribution. Because ore grades are declining generally, this has inevitably meant that higher tonnages have had to be treated. The increasing use of beneficiation at fine size for iron ores has provided a great stimulus for the development of large-scale, low-cost milling methods. An additional factor has been that in many cases the texture of the ore has become finer so that even more grinding has to be done to achieve liberation.

In the past, the problem of dealing with large tonnages was dealt with by the installation of a large number of what would now be regarded as moderately-sized ball and rod mills. The Nchanga concentrator in Zambia[4] is typical of this type of installation in which grinding was accomplished in eleven 9 ft × 8 ft (2.75 × 2.45 m) ball mills each handling 48 tonnes/h of new feed.

Today there are several solutions possible. These are (1) the use of very much larger ball mills either in combination with rod mills or on their own, (2) the use of autogenous and semi-autogenous milling techniques (3) the use of multi-stage and combination grinding circuits and (4) the use of specialised milling techniques.

The use of larger ball mills on their own can scarcely be better exemplified than in the case of the Panguna concentrator at Bougainville[5], which in 1972 was treating 78000 tonnes/day in eight primary ball mills of 18 ft

Table 1: Operating results at Bougainville for first six months of 1972

Total value at mine*	$37,664,720.00
Value per ton	$6.23
Operating costs	$17,730,900.00
Cost per ton	$2.93
Depreciation—Amortization	$5,942,538.00
Cost per ton	$0.98
Interest	$4,224,372.00
Cost per ton	$0.70
Total costs	$27,897,810.00
Cost per ton	$4.61
Net earnings	$9,766,910.00
Net per ton	$1.62

* Copper price for three months 1 April to 30 June was $0.479 (London), and $55.54 per ounce of gold.

Table 2: Production at Bougainville April 1 to July 1 1972

	Ore	Concentrate
	(tons produced)	
Tons milled	6,044,500	122,690
Average tons per calendar day*	60,929	1,348
Percent copper	0.67	27.3
Total copper, pounds	80,996,300	66,988,740
Ounce gold per ton	0.0215	0.806
Total gold	129,957	98,888
Ounce silver per ton	0.072	2.49
Total silver	435,204	305,498
Copper recovery†	82.70%	
Gold recovery†	76.09%	
Silver recovery†	70.20%	

* Remember shut down times including three typhoon days.

† Recovery was low because of high near surface oxide copper content.

dia × 21 ft (5.5 × 6.4 m). Each mill is fed with 406 tonnes/h, grinding to 45% −200 mesh (20–25% +65 mesh Tyler). Classification is carried out in eight 20 inch (50.8 cm) cyclones.

Operating at this scale gives a cost breakdown as shown in Table 1. During the period covered by the cost analysis above 9,592,500 tonnes of ore were milled producing 169,915 tonnes of concentrate, ie 0.018 tonnes of concentrate per tonne milled. The total statistics for a 3-month period in 1972 are shown in Table 2.

This type of grinding using a single large ball mill is being incorporated into the design of the Metcalf plant of Phelps Dodge and the Pinto Valley plant of Cities Service Co. The technique generally means lower capital costs and simpler operation than the conventional rod plus ball mill configuration. It is also employed on a smaller scale in many other plants.

An alternative solution to the problem is to resort to autogenous and semi-autogenous grinding in order to produce finished size product ready to enter the concentration process. Autogenous grinding is accomplished in large diameter, short length mills fed with run-of-mine ore. The diameter: length ratio of the mills is between 2 and 3 to 1, recent practice favouring the 2: 1 ratio.

In pure autogenous milling the ore acts as its own grinding media, in semi-autogenous milling a small proportion of steel balls is added. Achieving true autogenous milling is not easy as several important requirements must be met: (1) the mill must be large enough to lift the ore to a suitable height so that sufficient energy is developed for crushing (2) the ore fragments entering the mill must be of sufficient size to accomplish crushing (3) and most importantly, the ore must be hard enough so that there is always an adequate supply of correctly sized particles in the mill to achieve the correct grind. This latter point is very difficult to realise in practice and very difficult to predict without carrying out reasonably large-scale tests. It is also the point where

Table 3: Some autogenous and semi-autogenous mills used in the copper industry

COMPANY	No of Mills	Mill Size	HP
SOMINA, Mauritania	1 (dry)	24½ft × 4½ft	1500
MTD, Mangula	2 (dry)	23½ft × 5ft	1250
North Coldstream Mines, Canada	1 (dry)	18½ft × 5ft	600
Messina Development, Transvaal	1 (dry)	16½ft × 4ft	N.A.
GECAMINES, Zaire	2 (wet)	28ft × 10ft	3500
Lornex Mining, Canada	2 (wet)	32ft × 15ft	8000
Similkameen Mining, Canada	3 (wet)	32ft × 14ft	8000
Island Copper, Canada	6 (wet)	32ft × 14ft	7000
Pima Mining, Arizona	2 (wet)	28ft × 12ft	6000
UV Industries, Arizona	1 (wet)	22ft × 7ft	1500
UV Industries, Arizona	1 (wet)	26ft × 10ft	3000
Black Sea Copper, Turkey	3 (wet)	26ft × 10ft	3300
Cobar Mines, Australia	2 (wet)	22ft × 7ft	1500
Kanmantoo Mines, Australia	1 (wet)	22ft × 7ft	1250

some installations fail, hence the need to resort to semi-autogenous milling, which is a rapidly growing practice.

Table 3 shows a selection of semi-autogenous and autogenous mills used in the copper industry[6].

The Island Copper plant[7] at Vancouver Island, B.C., was constructed with autogenous mills, six 32 ft × 14 ft (9.75 × 4.26 m) mills were installed, but the plant is now being modified by the addition of a secondary ball-milling stage due, no doubt, to the problem of obtaining the right grinding charge size distribution.

Comparing iron ore and copper ore milling practices in which larger tonnages are treated than with most other ores, it may be said on the whole that in copper concentration true autogenous milling has not been so successful as is the case with iron ores, although even here, the success has been limited.

A good example of semi-autogenous milling practice is at the Similkameen copper concentrator in Canada[8], which treats 15,000 tonnes per day of 0.55% copper ore, grinding to 65% −200 mesh in three 32 ft × 14 ft (9.75 × 4.26 m) mills revolving at 10.4 rpm (76% critical). Each mill has two 4,000 hp motors and it is found that, by adding 7% of the total mill charge of 4 inch (10 cm) steel balls, double the output over that achieved by normal autogenous milling was realised. Consumption of balls is now 1.4 lbs/ton (635 g/tonne) and linear consumption 0.15 lb/ton (68 g/tonne)[9]. No detailed costs are available for this operation.

Multi-stage grinding is yet another solution to the problem, where the object is to remove from the circuit as often as possible material which has been ground to size. This type of installation can be seen at the Highland Valley, B.C., concentrator operated by the Bethlehem Copper Corporation[10]. This uses four stages of milling.

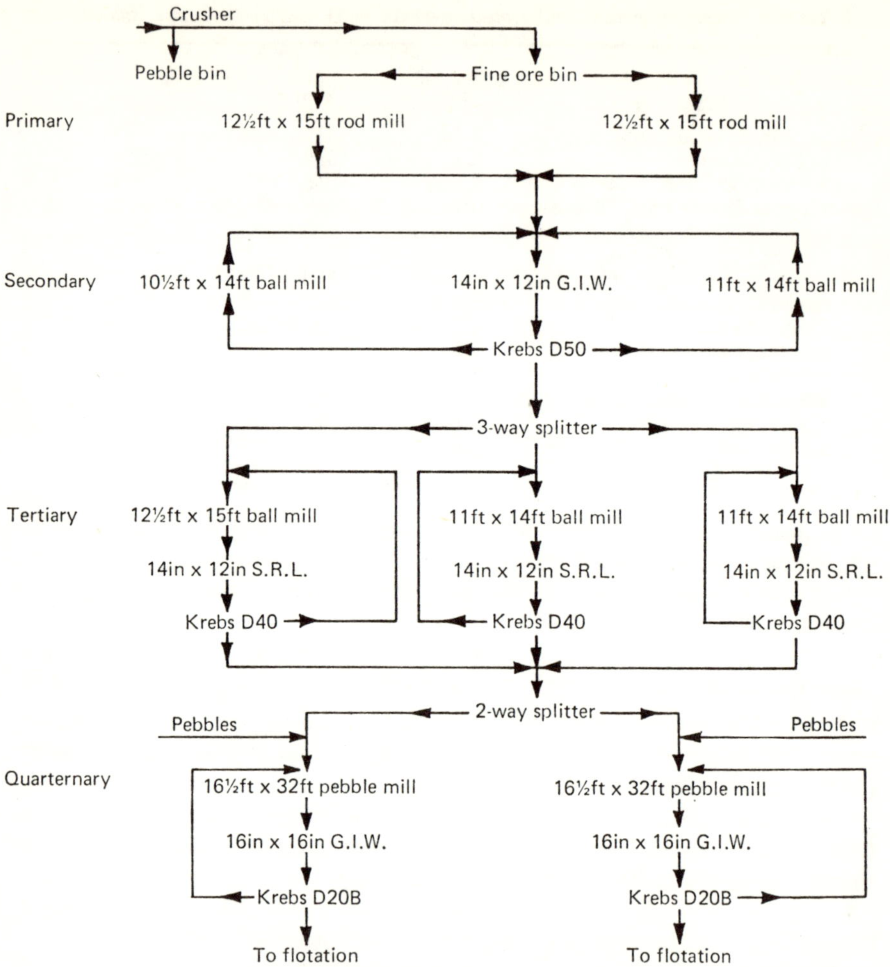

Fig. 6. Milling circuit at Bethlehem

The circuit layout is shown in Fig. 6.

At the Boliden Co.'s Vassbo concentrator[11], treating 600 tonnes/day of lead-zinc ore, the primary circuit has been altered from a rod mill to a 22 ft × 7 ft (6.7 × 2.13 m) Cascade mill in closed circuit with a hydrocyclone before the ball and pebble milling stage. The grinding statistics for this circuit are

	Primary	Secondary
Energy requirement kWh/ton	23	6.5
Grind size $\% - 44\ \mu m$	46	63
Production of $-44\ \mu m$ per kWh	20	26

This illustrates that beneficial changes can be made to plants having quite modest throughputs.

In the example which follows, the tests were carried out at the laboratories of Pickands Mather in the USA, in order to determine the best milling circuit for an iron ore which had to be ground to 80% −325 mesh Tyler[12].

The conventional rod plus ball mill in closed circuit with a cyclone was taken as the standard circuit and others were compared with it.

For the autogenous primary grind plus secondary ball milling, 16% lower capital and 15% lower operating costs were indicated. Primary autogenous milling followed by pebble milling, in which the pebbles were derived from the autogenous mill, indicated that lower operating costs would be incurred but capital costs were similar to the standard circuit. The operating cost pay-back time was longer also than that for the autogenous plus ball mill circuit.

In a circuit consisting of autogenous plus ball mills in which pebbles from the autogenous mill were removed, crushed and fed back as required, a 20% increase in feed rate was obtained when feeding back 5–10% of the feed. Moreover, output size from the mill could be controlled by varying the sizing and amount crushed. This enabled a good balance to be obtained with the tonnage fed to the ball mills, so that they could run at maximum efficiency. This circuit gave the lowest capital cost of all and a similar circuit was installed at Similkameen[8]. The simplest flowsheet of all, a single autogenous mill in closed circuit with a cyclone gave the lowest costs of all. When the plant was eventually installed at Wabush in 1965 it used six 24 ft × 18 ft (7.31 × 2.43 m) mills each powered by 1750 hp motors. The savings in capital and operating costs over a standard rod plus ball mill circuit were calculated to be $2 million in capital costs and $1 million in operating costs per annum.

It is of interest to note that the development work extended over 18 months, involved over 90 milling tests and cost $250,000; this underlines the value of adequate testwork, the cost of which will be recovered many times over.

The use of pebble grinding in which no steel media are used in the secondary milling stage is spreading. The pebbles are derived either from the crushing stages or from autogenous milling in the primary stages. Øyasaeter[13] lists data obtained from eight mills in S. Africa, eight in Finland, eight in Sweden and eight in the USA and Canada. In pebble milling there is no cost for steel grinding media, though some cost is incurred in obtaining, crushing, screening and feeding the pebbles. In an existing mill the change over from ball milling to pebble milling does not involve such major changes as would be the case if the change was made from rod milling to autogenous milling, and adaptations can prove very successful. Again, adequate testwork is necessary to determine the viability of such a change, in particular that an adequate supply of pebbles will be assured. Suitable warnings of problems in pebble milling are outlined in Øyasaeter's papers.

Mention must now be made of an interesting development which has recently been announced after some extensive trials have taken place. This comes under the category of specialised milling techniques and is the Erwo milling system developed in Austria jointly by Waagner-Biro A G, the Austrian-American Magnesite Company and Chemie und Metall GmbH[14]. The system consists of a mill with a special liner design which may be combined with a special design of classifier, a hydrocyclone and a Lamella thickener. In some applications, e.g. in grinding cement clinker the mill only would be required. The object of the system is to reduce power consumption, increase throughput and produce a closely size-ranged product with minimum slime loss.

The mill is fitted with special liners which give the mill at any given point the cross section of a rounded off square. In each succeeding section the corners of the squares are set in a spiral. This is termed an angular spiral lining (asl) and is claimed to increase impact grinding and lessen abrasion grinding by giving a better throw to the mill contents, also optimum grinding efficiency is reached with lower media charges than normal, less power is consumed and higher throughputs are achieved.

The classifier used with the mill is called the Bathmos classifier and is a multi-stage horizontal classifier operating on free-settling principles. The circuit usually employed is as follows: mill output is pumped to a hydrocyclone, the sands proceed to the Bathmos classifier where classifier sands are returned to the mill, cyclone and classifier overflows are fed to the Lamella thickener where they are thickened to 25% solids before being passed on for processing. Since the mill feed is free from undersize the circulating load can be reduced by 20 to 50%. Comparison tests using Rio Tinto ore sized 1% +8 mm (0.3 inch) with 12% −0.025 mm (0.001 inch), using conventional and Erwo circuits using a 35 ft^3 (1 m^3) mill fed with 72% solids and running at 70% critical, gave the results shown in Table IV.

Table 4.

Stage	No	Mills Size	Type	Type of Circuit	Installed HP
Primary	2	12½ × 15ft (3.8 × 4.6 m)	Rod	Open	5,600
Secondary	2	11 × 14ft (3.4 × 4.3 m)	Ball	Closed	4,100
	(2	11 × 14ft (3.4 × 4.3 m))			
Tertiary	(1	12½ × 15ft (3.8 × 4.6 m))	Ball	Closed	7,700
Quarternary	2	16½ × 32ft (5.0 × 9.75 m)	Pebble	Closed	11,000
					28,400

The Erwo circuit gave
4% more product for flotation
20% less overground −10 μm material
50% less 0.4 mm oversize
16% less energy consumption
30% less circulating load
than the conventional circuit.

Additional data[15] show that grinding cement clinker in a mill fitted with asl, the following improvements shown in Table 6 were obtained. A 40 tonnes/h plant has been recently installed in Zambia.

Table 5: Comparative mill circuit results

	Conventional Circuit	Erwo Circuit
Flotation feed*	1,295 kilograms per hour solids at concentration of 25% by volume	1,350 kilograms per hour solids at concentration of 25% by volume
	0.4 millimetre—6% minus-0.01-millimetre-31%	0.4 millimetre-2.5% minus-0.01-millimetre-25%
Circulating load returned to ball mill	5,700 kilograms solids per hour with	4,000 kilogram solids per hour with
	1,100 kilograms per hour minus-0.1-millimetre	160 kilograms per hour minus-0.1-millimetre
Mill power consumption	14.9 kilowatts	12.5 kilowatts

* To protect the narrow apex nozzles of the hydrocyclones, a 3-millimetre (0.12-inch) protective screen was included in the circuit. The sand retained on it was rejected to waste.

Table 6: Effect of ASL lining on mill results

Item	Normal lining	ASL lining
Power consumption kW	770	660
Output tonnes/h	9.6	11.6
Output per annum, tonnes	70,080	84,680
Annual energy consumption kWh	5,620,416	4,809,824
Saving of energy kWh/tonne	—	23.4
Annual cost saving power cost in Austrian Schillings per annum	—	405,296
Increased output tons/yr		14,600

Finally mention should be made of the savings which can be made by the use of rubber linings for mills, a practice which was pioneered in Scandinavia and which is now spreading.

The conclusions reached after a study of Canadian practice[16] are that rubber liners last about twice as long as steel liners, they are easier and faster to install due to their lighter weight and noise level is reduced. Rubber discharge grates for mills block less easily, are easier to handle and wear better than their steel counterparts. However, rubber lifter bars wear faster than those of Ni-hard steel. At the Dickinson concentrator re-lining time for an 8 ft × 12 ft mill (2.44 × 3.66 m) was reduced by 60% and at Denison, grinding cost was reduced from 4.837 to 4.716 cents per tonne milled, a saving of approximately $53 per day.

Results from Russia[17] show that when using rubber linings in 7 ft × 10 ft (2.1 × 3.0 m) ball mills with 40–60 mm (1.56–2.36 inches) balls, lifter bars wore by 40–55% and front end bars wore by 8–10% after 18,500 hours running (about 2 years).

Pre-concentration

This is a field in which, if liberation of an ore is coarse enough, considerable upgrading of the ore can be made. This has several advantages; the subsequent circuit will probably perform more efficiently on enriched feed, the total plant will be able to deal with more input, keeping the load on the concentrating section the same, variations in head grade can be smoothed out and, finally, it is possible that some deleterious minerals present in the head ore may be at least partially discarded. It is in this context that an examination is made of two important pre-concentration techniques which can help to achieve the above objectives and save overall costs. The two topics dealt with are sorting and heavy media cyclones.

Sorting

Manual methods of sorting by "hand picking" have been practised for a very long time and have proved useful in achieving a satisfactory degree of waste rejection particularly in areas where labour cost was low and employment of local labour was important. The rise in the cost of labour and the larger tonnages which have to be processed have tended to militate against these simple techniques.

Derived from machines which were originally developed for sorting peas, beans etc, a whole family of sophisticated purpose-built machines has been developed which greatly extend the methods available to effect this separation. The object of sorting is to use some property of the lump ore or gangue to differentiate one from another so that the rich ore may be retained and the waste rejected. The properties which are available for this technique are varied and include diffuse reflectivity, colour, visible fluorescence in X-rays, fluorescence in ultra-violet light, opacity, translucence, infra-red iridescence and conductance. The size at which these separations can be carried out is generally in the range 3–170 mm (0.12 to 6¾ inch) with the proviso that sufficient liberation exists in this size range for the process to work efficiently. Basically the particles must be presented to the sorting machine in a stream which passes a sensor selected to detect the radiation chosen to discriminate between values and waste. A signal generated in the sensor is processed electronically which in turn actuates some device which directs the particles into appropriate values and waste chutes. The directing device is usually a high speed air valve. For a full explanation the paper by Balint[18] should be consulted. The layout of the Gunson 811 M, is illustrated in Fig. 7.

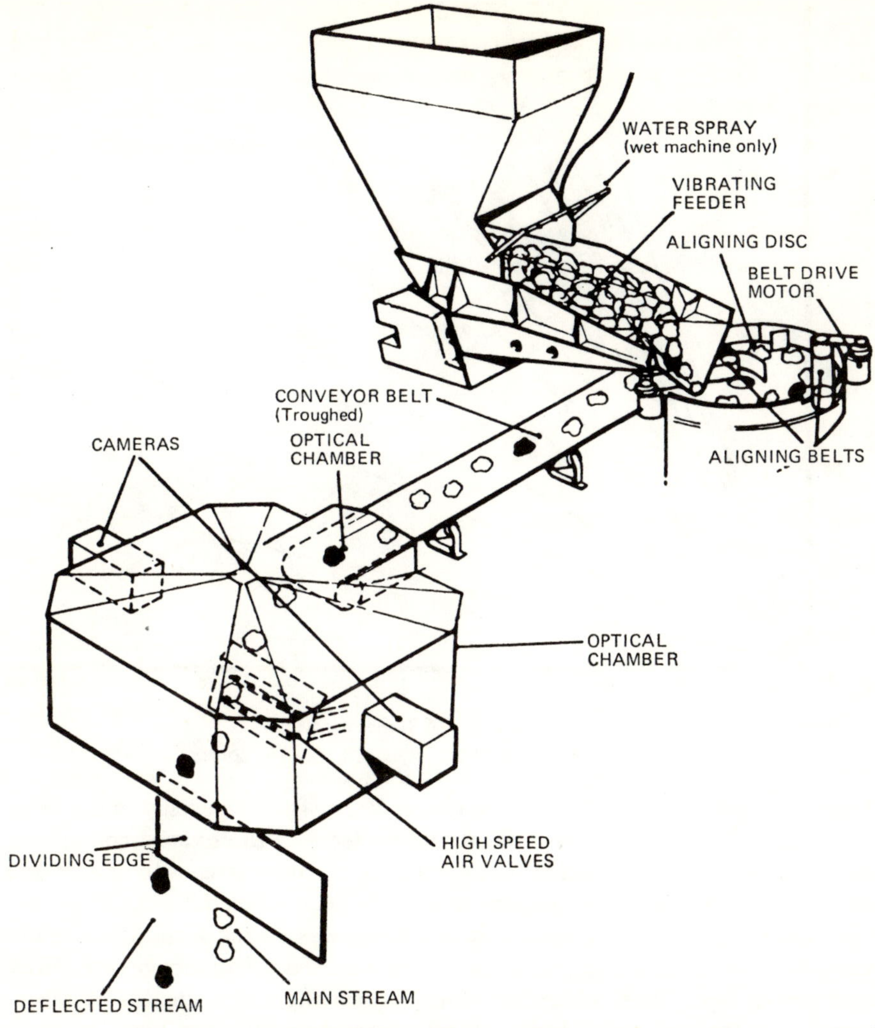

Fig. 7. Layout of Gunson 811M sorter for 50–150 mm particles

One of the main parameters which determines the cost of the sorting operation, apart from the capital cost of the equipment, is the rate at which the particles can be processed through the separator and the mass of each particle. The feed rate characteristics for the 811M optical sorter are shown in Fig. 8. It can be seen that the rate varies from just under 50 tonnes/h when sorting pieces weighing 2.75 kg each at the rate of just under 5 per second, to about 15 tonnes/h sorting 0.25 kg lumps at about 18 pieces per second. Note how sorting rate increases for the smaller-sized stones. It should be mentioned that the size range to be sorted through one machine can be in the ratio of 3:1 in the 50 to 150 mm (1.96 to 5.9 inch) size range, but should be only

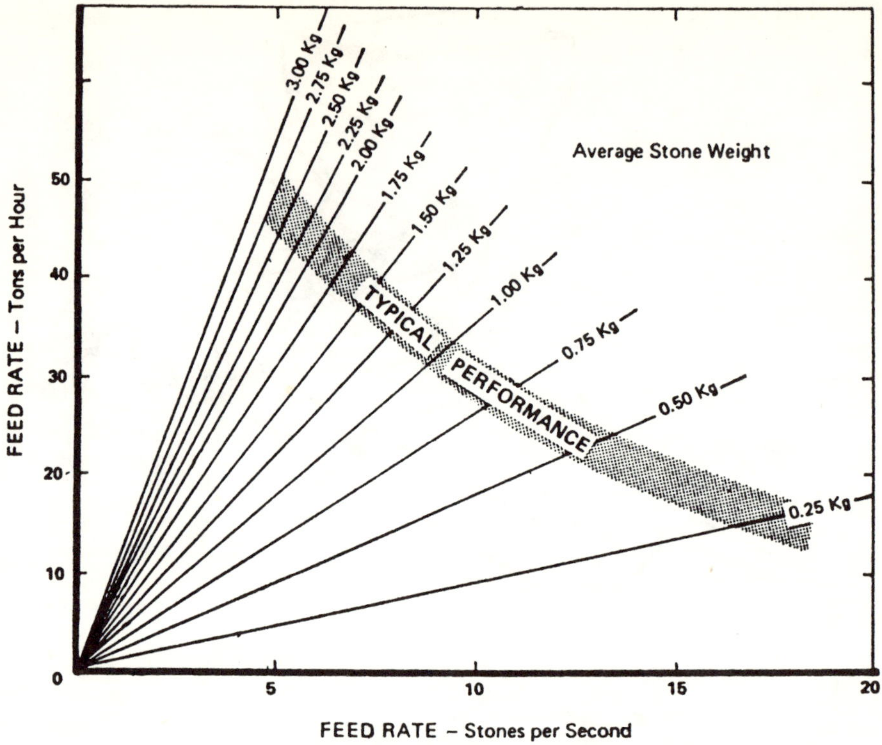

Fig. 8. Feed rate characteristics of the 811M sorter

1.5:1 in the 10 mm (0.39 inch) size range. Very often, especially in the large size range, the particles are washed and sorted wet to enhance colour and wash off dirt but in the smaller sizes water causes the particles to adhere and they are ideally sorted dry, having previously been washed.

Some typical applications include sorting waste rock from scheelite ore using ultra-violet fluorescence, sorting dark waste rock from magnesite, black and brown flints from white limestone, improving the whiteness of fillers, removing iron-stained dolomite from clean dolomite for glass making, separating contaminants from rock salt and sorting diamonds from gravel using X-ray fluorescence.

Sorting flint from limestone is illustrated in Table 7.

Table 7: Sorting flints from limestone

Product	Wt %	Composition Flint	Composition Limestone	% CaO	Distribution % Flint	Distribution % Limestone
Concentrate	83.1	1.6	98.4	94.1	12.0	91.6
Reject	16.9	56.4	43.6	42.8	88.0	8.4
Feed	100.0	10.9	89.1	85.4	100.0	100.0

The feed was sized 60 to 170 mm (2.36 to 6.69 inch) with an average weight of 1.64 kg per piece. Two 811M optical sorters sorted 88 tonnes/h giving a better product than the previous hand-sorting methods.

Some data, which include costs, are available from an operation in Finland sorting limestone for cement manufacture[19].

The limestone is traversed by various intrusions rich in silica whose colour ranges from pink to white but which are mostly grey. The limestone is white but ranges from pink to grey. Perfect separation is not possible as the limestone assays only 90–95% $CaCO_3$. The results of several trials are shown in Table VII.

Table 8: Separation results on a Sortex 811 M machine at Ojamo

	Test No: 47	Test No: 48	Test No: 49	Test No: 50	Test No: 51	Test No: 52	Test No: 53	Test No: 54	Test No: 55
Sample Weight, Tons	14.5	13.7	13.5	12.7	10.5	21.00	13.7	13.5	12.5
Average Stone Weight Kg per piece	0.56	0.68	0.62	0.78	0.61	0.61	0.54	0.60	0.49
Feed Rate, Tph	19.3	21.9	20.3	25.4	21.0	21.0	18.3	27.0	16.7
Concentrate Weight	89%	94%	86%	94%	74%	61%	93%	89%	79%
$CaCO_3$ Content									
Feed	77.5%	82.1%	73.5%	81.4%	67.1%	81.0%	74.4%	77.2%	78.3%
Concentrate	85.4%	86.0%	81.4%	85.6%	84.9%	89.2%	81.3%	85.9%	86.3%
Recovery of $CaCO_3$	98.8%	99.4%	98.5%	99.3%	96.4%	62.9%	99.6%	98.8%	96.9%
Machine Settings									
Filter	Red	Red	Red	Red	Red	Red	Red	Red	Red
Discriminator (Machine units)	6.5	7.0	6.6	7.0	6.8	6.5	6.7	6.7	6.3
Comparator	4	1	1	9	9	9	4	4	4

Feed: Normal stone wet

Slight changes were made to the feed sizes in actual operation, changing from 50–100 mm (1.96 to 3.93 inch) with a piece weight of 0.49 to 0.78 kg to 70–130 mm (2.75 to 5.1 inch) with a piece weight of 1.05 to 1.20 kg. This gave an increased feed rate of 40 tonnes/h on the 811M machine, and in plant operation, 95% of the $CaCO_3$ was recovered into a concentrate assaying 85–90% $CaCO_3$, which was blended with other material to give an overall mix suitable for cement manufacture. Typical operating costs for the plant in Finland in 1970 are given in Table 9.

As a guide to capital costs the cost of a separator for the 150–5 mm (5.9–0.2 inch) size range would vary from $2700 to $11,000 per ton of hourly capacity, the lower costs being for sorting large-sized lumps. Power consump-

Table 9: Operating costs for the Ojamo plant

	Costs per ton of feed Finnish marks	Equiv. US $
Power consumption including water pump compressor and external conveyor belts	0.06	0.014
Operators	0.20	0.048
Maintenance	0.03	0.008
Total	0.29	0.070

The capital cost breakdown was in terms of percentage

Cost of 811M separator	54
Supporting structure and materials handling equipment	38
Compressor	8
	100

tion varies from 0.3 to 0.5 kWh per ton of feed and air consumption is about 56 m^3 (1977 ft^3) per ton of material ejected, ie if 10 tons per hour are treated and 30% is ejected, air requirement is 0.3×56 m^3 or 2.8 m^3/min (100 ft^3/min) of free air, which is usually used at 5–6 atmospheres pressure. Maintenance is usually about 5–20 cents/ton including labour, depending on the care taken of the plant.

Heavy Media Cyclones

The heavy media cyclone was developed to enable the range of sizes which can be treated by conventional drum and cone sink-float methods to be extended downwards from the bottom limit of $\frac{3}{8}$ inch (9.5 mm) to about 36 mesh (0.42 mm). In the normal sink-float separations carried out in drums, cones or baths the process is carried out in relatively slow-moving suspensions of fine ferrosilicon, magnetite or mixtures of the two maintained at a density which will enable the desired separation to take place. In the heavy media cyclone, the linear velocities of the media and pulp are much higher in order to achieve the enhanced gravity conditions. Generally, although introducing certain operational problems, this leads to high throughputs for smaller units so that the machine has found extensive use in coal cleaning applications and has also had considerable impact in other branches of the mineral industry.

Before considering performance it is important to note that the heavy media cyclone itself is only a small part of the total plant involved in carrying out this type of separation. Fig. 9 illustrates the various stages in a typical heavy media cyclone plant, six stages are involved which cover feed prepara-

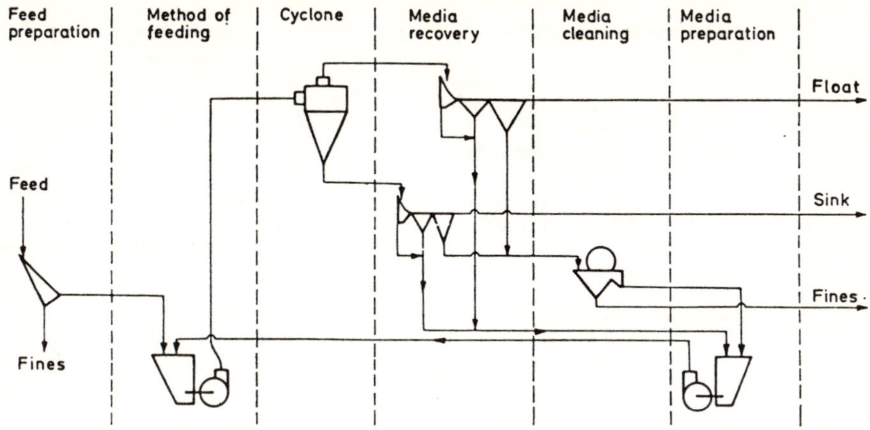

Fig. 9. Typical heavy media cyclone flowsheet

tion, feeding, cycloning, separation, cleaning and densifying the media and dealing with the products. Many different types of heavy media cyclone have been developed, these include the Univer, Dyna Whirlpool, Dutch State Mines cyclone, and the Vorsyl separator.

The following results pertain to some tests carried out using the Univer heavy media cyclone on a magnesite ore[20]. The Univer is essentially a cylindrical cyclone in which both the streams carrying sinks and floats are pumped out of the cyclone, which is different from the normal cyclone in which feed is pumped in, the exit flows being controlled by orifice sizes. The tests were carried out in a 10 inch dia (25.4) cm pilot-scale plant using a magnesite head material assaying 1.26% SiO_2, 1.96 CaO and sized 4–12 mm (0.15 to 0.47 inch). Feed was passed over a 0.5 mm (0.19 inch) screen and washed with sprays before being fed to the Univer, the medium used was Knapsack cyclone 40 which was sized 90% −325 mesh Tyler (−43 μm).

The conditions for the test were

Floats media Sp Gr	2.24	Floats product % wt	15.3
Sinks media Sp Gr	3.00	Sinks product % wt	84.7
Media mixture Sp Gr	2.70	Cut point Sp Gr	2.915
Floats pump speed	1220 rpm	Ep	0.048
Sinks pump speed	850 rpm		

The analysis of the products carried out in heavy liquids is shown in Table 10. The economic data relating to the separation of 900 tonnes/day of feed sized 10 to 1 mm (0.4 to 0.04 inch) using an 18 inch (45.7 cm) dia unit are as follows:

Capital Costs			£
Feed arrangements			36 000
Plant			410 000
includes major items such as	Feed preparation	20 000	
	Handling fine rejects	44 000	
	Separator	50 000	
	Instrumentation	22 000	
	Sampling	2 000	
	Media handling	1 500	
	Product handling	40 000	

Operating costs	£ per annum
Feed arrangements (a) Power	2 000
(b) Spares	4 500
HMS unit (a) Power	8 000
(b) Labour	7 000
(c) Medium at 2–3 lb/tonne	15 500
Total	37 000

Depreciation and interest rate

Calculated from $K = \frac{A}{100}\left[\frac{100}{n}+\frac{P(n+1)}{2n}\right]$

where A = capital investment, £446 000
n = service life, say 10 years
P = annual interest rate, 15% — £81 395

Thus capital costs are £81 395 p.a. and operating costs are £37 000 p.a. Total capital + operating costs are £118 395 p.a.

Assuming 330 operating days per annum this gives a throughput of 270 000 tonnes p.a., and a capital cost of £0.301/tonne feed and an operating cost of £0.137/tonne feed. These costs are based on UK operations as at December 1971.

Table 10: Analysis of sink and float products from magnesite separation

	FLOATS		SINKS			
	Wt.%	Wt.% of feed	Wt.%	Wt.% of feed	Reconstituted feed%	Dist. Coeff.
Float at 2.85	3.77	0.58	0.06	0.05	0.63	7.94
2.85–2.90	7.63	1.16	0.62	0.53	1.69	31.36
2.90–2.95	6.88	1.05	1.80	1.52	2.57	59.14
2.95–2.975	15.23	2.33	3.72	3.15	5.48	57.48
2.975–3.00	42.14	6.45	37.37	31.65	38.10	83.07
Sink at 3.00	24.35	3.73	56.43	47.80	51.53	92.76
	100.00	15.30	100.00	84.70	100.00	

A rather different operation, but one which is being carried out on a very large scale, is the concentration of diamond gravels in S. Africa. Data from a paper by Chaston[21] indicate the aggregate hourly capacity of the plants presently installed exceeds 2,300 tonnes. The details are shown in Table 11.

The important features to note are the very high hourly treatment rates

Table 11: Operating details of some of the major diamond heavy media plants

Mine	Year installed	Feed to cyclones, ton/h	Feed pressure or head	Feed size range, mm	Specific gravity media			Media	Conc. ratio	Media usage, g/ton	Diamond recovery %
					Feed	Spigot	Overflow				
Williamson	1955	155	Pump 290 kPa	−6 +1	2.05	2.8	2.0	80% magnetite-20% FeSi	1:475	1600	97
Premier no 9 unit	1964	160–200	Pump 250 kPa	−6 +3	2.30	2.57	2.14	65 D FeSi	1:250	330	97
Dreyers Pan	1966	21	Pump 110 kPa	−20 +2	2.3	2.9	2.03	100 D FeSi	1:50	600	99
CDM No 1 plant	1968	80	Gravity 8.0 m	−25 +2	2.76	3.0	2.7	100 D FeSi	1:65	450	99
CDM No 4 plant	1969	360	Gravity 8.0 m	−25 +2	2.72	3.0	2.6	100 D FeSi	1:50	400	99
Premier retreatment	1970	340	Pump 210 kPa	−8 +1.8	2.62	2.95	2.56	65 D FeSi	1:125	350	98
Orapa	1971	350	Gravity 8.0 m	−25 +1.6	2.55	2.80	2.4	100 D FeSi	1:100	350	98
Koffiefontein	1971	560	Gravity 8.0 m	−30 +0.5	2.71	2.82	2.66	100 D FeSi	1:300	700	98
Kleinzee	1972	250	Gravity 8.0 m	−25 +2	2.68	2.96	2.60	100 D FeSi	1:75	Not yet available	99

Table 12: Operating costs of heavy media plants in S. Africa in South African cents per ton of feed (200 c = 2R = £1)

	No 4 plant CDM	Premier retreatment	Koffiefontein	Orapa	Williamson
Labour and supervision	1.1	2.5	2.0	1.9	1.3
Screening	1.6	1.5	1.4	1.9	
Media	3.4	2.8	6.8	5.4	5.3
Cyclone parts	0.4	0.2	0.1	0.1	
Pump parts	0.8	0.5	0.2	0.4	
Maintenance	1.6	2.7	1.5	2.6	2.6
Power and water	4.0	3.8	3.5	6.6	6.2
Overheads and sundries	2.3	2.8	2.1	1.5	1.5
Total	15.2	16.8	17.6	20.4	16.9
Ton/h to cyclones	360	340	560	350	155

and low media losses, especially in the latest plants and the high ratio of concentration necessary for this type of operation. The valuable nature of the final product coupled with the very low grade of the feed means that each unit operation in the process must be made very efficient especially the cyclone operation itself so that high recoveries can be achieved.

The costs of operating these types of plants are given in Table 12. The cost of media in S. Africa is also given by Chaston, and is shown in Table 13.

It will be seen from Tables 12 and 13 that the two most popular types of heavy media used in this operation were the milled Amcor grades 65D and 100D, costing 112 and 124 R respectively.

It seems that some of the advantages claimed for the more expensive grades of atomised ferrosilicon, such as less corrosion and less wear to due the rounded nature of the particles, have neither been proved or disproved in S. African practice and since the specific gravities reached are not particularly high with consequent viscosity problems the milled grades appear to be quite adequate.

Gravity Concentration

The main trends in gravity concentration have been firstly in developing methods for concentrating fine dense minerals, secondly in developing methods for concentrating minerals in the dry state and lastly in improving the efficiency of existing machines.

One of the most interesting of the new machines to be established is the Bartles-Mozley concentrator[22] a device for concentrating fine heavy minerals in a flowing film. The concentrator consists of 40 thin, flat fibreglass decks stacked over one another at a spacing of $\frac{1}{2}$ inch (1.3 cm). Each deck measures 4 ft broad by 5 ft long (1.22 × 1.5 m). The decks are located in a frame in two packs of 20, separated by a space in which an out-of-balance weight rotates in a horizontal plane at about 230–250 rpm, imparting an oscillatory motion to the deck packs. The frame is freely suspended on wires and in the running position is inclined downwards from feed to discharge end at an angle of between 1 and 3°. Pulp is distributed to all 40 decks at a total flow of 6–7 l/s (80–93 gal/h) at about 10% solids, for a period of up to 36 minutes according to the amount of material to be collected on the decks. The tailings are discharged continuously while the heavy fraction is held on the deck surface. After the decks are loaded, the feed and oscillation are stopped, the decks tilted to about 45° and the concentrate washed off into a separate launder. Fig. 10 shows the concentrator in the running and discharge positions. The

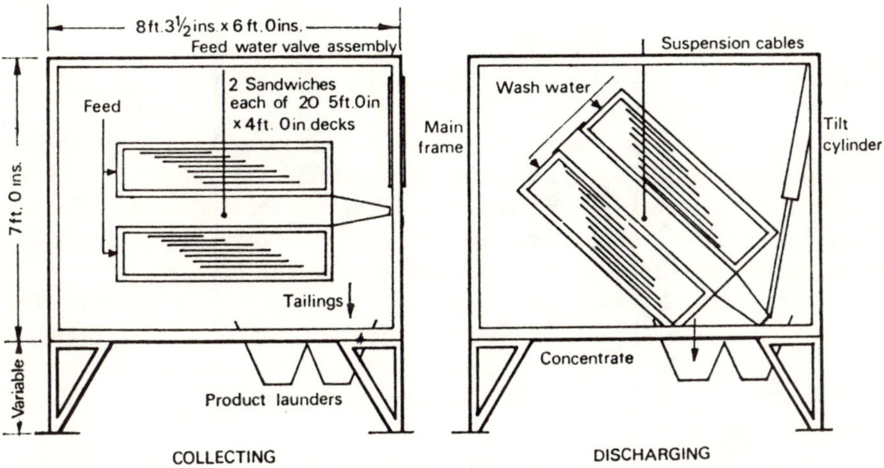

Fig. 10. The Bartles-Mozley concentrator—diagrammatic sketch

sequence of feeding and discharging is controlled by an automatic timer. In the washing operation wash water is applied at the rate of 160 gal/min (12 l/s) for 30 seconds per cycle giving a total water demand of about 80 gal/h (6 l/s).

The concentrator is designed to treat particles in the size range 5–100 μm for most normal minerals but heavier materials such as gold, silver and platinum could be recovered at even finer sizes. The concentrator is especially suitable for the recovery of fine particles of cassiterite, wolframite and tantalite in ores or tailings.

It is estimated that direct operating and maintenance costs should not exceed £0.03 per tonne of feed. The capital cost of the machine is now £5,400 ex works, ie approximately £2,200 per tonne of hourly capacity. A slime table by comparison would cost in excess of £4–6000 per tonne of hourly capacity depending on feed characteristics. The range of applications for the concentrator is summarised in Table 14. A series of 45 Bartles-Mozley tables

Table 13: Cost of media in S. Africa

Name	65D	100D	150D	270D	Normal fine	Cyclone 60	Cyclone 40	200 XD
Manufacturer	Amcor	Amcor	Amcor	Amcor	Amcor and Hoechst	Amcor and Hoechst	Hoechst	Foote Mineral Co
Type	Milled	Milled	Milled	Milled	Atomized	Atomized	Atomized	Milled
Size, Tyler	Wt%	Wt%	Wt%	Wt%	Wt%	Wt%	Wt%	Wt%
+65	1	—	—	—	1	—	—	—
−65+100	2	—	—	—	7	—	—	—
−100+150	5	1	1	—	10	2	—	—
−150+200	12	4	1	—	15	5	2	12
−200+325	35	30	23	10	22	20	8	28
−325	45	65	75	90	45	73	90	60
Si, %						All 14–16%		
Sp. gr.						All 6.7–6.9		
Price, Rand/tonne in South Africa	100	112	123	145	173	306	335	206

Table 14: Summary of results for materials tested on the Bartles-Mozley concentrator

Sample No	Material and Size	Density of Economical Mineral	Through-put MTPH	% Assays Feed	% Assays Conc.	% Re-covery	Enrich-ment Ratio
1	Wolframite tailing deslimed—150 mesh	7.2/7.5	2.5	0.06% WO_3	0.26% WO_3	88	4.3
2	Cassiterite-pyrite ore 95%—300 mesh	6.8/7.1	1.6	0.83% Sn	2.40% Sn	74	2.9
3	Cassiterite ore 30%—10 microns	6.8/7.1	2.6	0.5% Sn	1.17% Sn	63	2.3
4	Cassiterite fines 64%—10 microns	6.8/7.1	2.0	0.36% Sn	2.22% Sn	46	6.2
5	Tantalite tailing 95%—200 mesh	5.3/7.3	2.2	0.10% Ta_2O_5	0.44% Ta_2O_5	48	4.4
6	Scheelite tailing 95%—200 mesh	5.9/6.1	N/A	0.22% WO_3	0.62% WO_3	69	2.8
7	Cassiterite tailing 95%—200 mesh	6.8/7.1	3.5	0.38% Sn	0.81% Sn	71	2.1

has been installed at the Ecstall Mining Co's mill at Timmins, Ontario, to recover cassiterite from the copper flotation tailing. The flowsheet adopted can be seen in Fig. 11.

The 45 concentrators work in conjunction with 32 Bartles-CTS screens (see earlier). Pilot-plant tests indicate that from a head of 0.18% Sn, a concentrate of 15.4% Wt can be recovered at a grade of 0.7% Sn for a recovery of 61.2% of the tin. The Bartles-Mozley concentrates amounting to 500–550 tonnes/day are then upgraded to 27 to 28% Sn on Holman slime tables which make a stage recovery of 75%. The Holman concentrate is further upgraded by flotation to remove the pyrite and finally acid-leached to remove iron minerals.

In the dry-processing field, two devices have been developed, the first, an air-fluidised pinched sluice[23] and the second a dry heavy-media separator using air-fluidised heavy medium[24].

The sluice is a high-throughput device for concentrating sand-sized feeds, a schematic diagram of the sluice is shown in Fig. 12. The sluice takes the form of a shallow, tapered trough fitted with a porous plastic deck. An air box integral with the trough permits low pressure fluidising air to pass upwards through the deck. At the end of the trough an adjustable splitter unit devides the material stream into "heavies" and "lights". The sizes which can be treated range from 10 to 200 mesh and ideally specific gravity differences of 1.0 are required to effect satisfactory separations between the components of a mixture.

The main advantages of the process are low running costs, no moving parts and high throughput. Table 15 lists the technical specifications of the two standard models available. The sluice can be used either as a concentrator, a pre-concentrator or as a classifier when the feed is mono-mineralic.

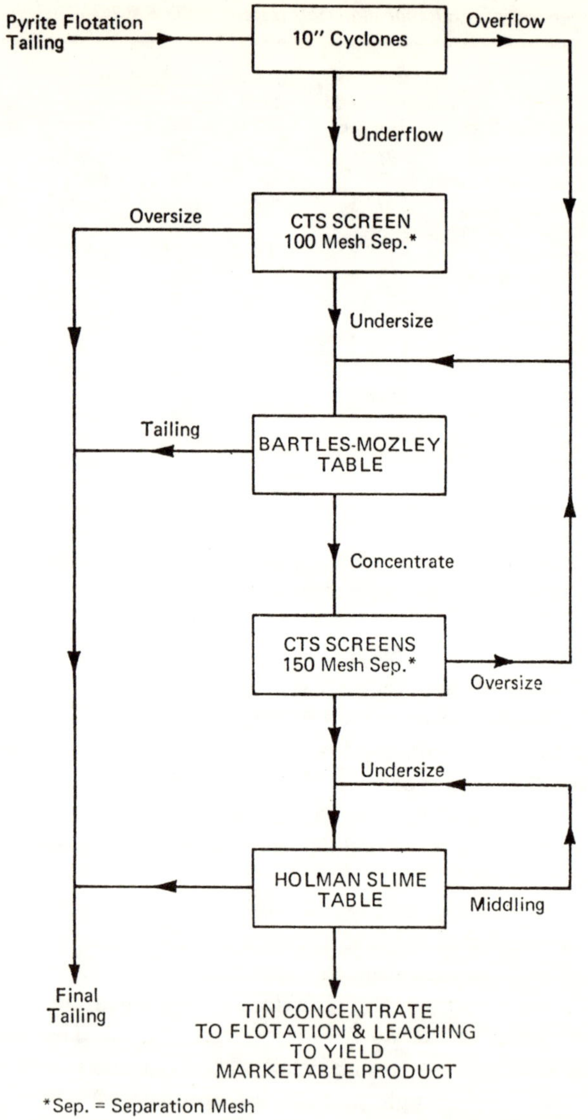

Fig. 11. Flowsheet for treatment of copper tailings at Timmins

An example of the sluice used as a concentrator or pre-concentrator is shown in Table 16. The feed, a tin pegmatite with quartz, feldspar and mica as principal gangue minerals, had a wide size range, 80% by weight falling between 25 and 150 mesh. A feed rate of 2050 lb/h (929 kg/h) was achieved on a 30 inch (76 cm) sluice. Air flowrate was 1800 $ft^3/h/ft^2$ (548 $m^3/h/m^2$). Acting as a concentrator, the sluice recovered 65% of the Sn at a grade of

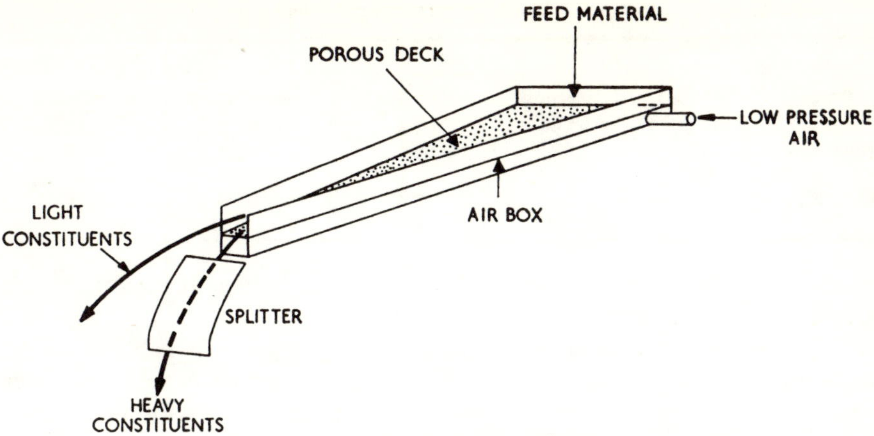

Fig. 12. Air fluidised pinched sluice

Table 15: Sluice technical specifications

	MODEL 30	MODEL 60
Trough length	30 inches	60 inches
Trough width (feed)	9 inches	18 inches
Trough width (discharge)	1 inch	2 inches
Overall length	5 feet 8 inches	8 feet 6 inches
Overall width	2 feet	3 feet
Height	3 feet 6 inches	3 feet 6 inches
Fan capacity	150 cfm at 10 inch wg*	150 cfm at 10 inch wg
H.P. (Fan)	¾ h.p.	¾ h.p.
Capacity	¾–2 tons/hour	3–8 tons/hour
Bed area	1 square foot	4 square feet

* The fan is rated so that more than one separator can be run from one unit.

Table 16: Sluice performance on a tin pegmatite

Product	% Wt	Assay % Sn	% Dist	Cum. % Grade	Cum. Dist %
Concentrate	3.3	54.0	65.1	54.0	65.1
Middlings	14.7	4.6	24.7	13.7	89.8
Tailings	82.0	0.34	10.2	2.73	100.0
Calc Hd	100.0	2.74	100.0	—	—

54%. If the middlings were combined with the concentrate recovery increased to nearly 90% at a grade of 13.7%.

In order to assist with difficult separations and to help fluidisation, heavy media may also be used with the sluice, which is screened out from the products and re-used. If separations are involved in which very heavy materials

Table 17: Results of classifying building sand

Product	Wt %	Content Wt% +100 mesh	Content Wt% −100 mesh	Distribution % +100 mesh	Distribution % −100 mesh
Coarse	93.9	92.9	7.8	97.7	64.1
Fine	6.1	32.8	67.2	2.3	35.9
Calc Hd	100.0	88.6	11.4	100.0	100.0

have to be treated such as lead, steel shot or metal fragments then the sluice may be vibrated to assist fluidisation.

An example of the sluice used as a classifier is given in Table 17. A 30 inch (76.2 cm) sluice was used to classify a building sand, sized −14 mesh with 12% −100 mesh. Feed rate was 840 lb/h (380 kg/h) with an air consumption of 2800 $ft^3/h/ft^2$ (853 $m^3/h/m^2$). The applications of the machine are numerous, ranging from the concentration of heavy minerals from beach sands, of tin, copper and iron ores and phosphate rock, the separation of shot from foundry sands, of steel grindings from abrasive, of wire from plastic (using media) and classification of sulphur, limestone, gypsum, burnt lime, dolomite etc. The capital cost of a 60 inch (152 cm) model is £3–4,000 giving a specific cost varying from £1200 to as low as £440 per ton of hourly capacity. Installed power is ¾ hp for each model. It is a useful machine for work in arid areas, or where a dry process is specified or the material must not be wetted.

Acting in a complementary role to the sluice is a dry separator designed for sorting lumps ranging from 2 or 3 inches (5 to 7.6 cm) down to about 22 mesh using an air-fluidised medium such as fine ferrosilicon to effect a sink float separation in a similar fashion to that used in current bath-type heavy media plants. A diagram showing the general arrangement of a separator is shown in Fig. 13.

The full scale separator has a 4 ft (122 cm) wide bed and contains its

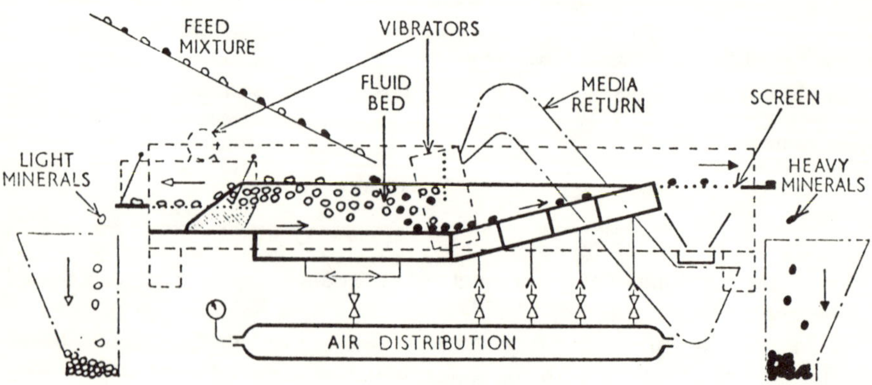

Fig. 13. Fluid bed separator

own magnetic media cleaning circuit, it has a throughput of 5 to 10 tonnes/h depending on the density and size range of the feed and ease of separation. Probes in the fluidised bed measure the bed density continuously and this can be linked to the media cleaning circuit to control bed density. Similar separations to those achieved with normal dense media are possible and media losses are similar provided the feed is perfectly dry. Capital cost of a 4 ft (122 cm) machine is about £16,000 and includes all media-cleaning equipment, blowers, feeders and control panel. The installed hp for the machine is 9 giving a specific installed horse power of 0.9 hp per ton of throughput under the most favourable conditions.

An example of the separations possible is given in Table 18 which refers to a two-stage separation of a barytes, fluorite, galena ore at bed densities of 2.78 g/ml and 3.3 g/ml. The feed was sized $-\frac{1}{4}$ +22 mesh. The resulting 3 products were separated in heavy liquids at densities of 2.78, 3.3 and 4.5 g/ml with the results shown in Table 18.

Table 18: Metallurgical balance for fluorite-barytes-galena separation

		Fluorite		Barytes		Galena	
Product	Wt. %	Assay % CAF_2	Distribution %	Assay % $BaSO_4$	Distribution %	Assay % Pb	Distribution %
Float at sp. gr. 2.78	25.1	12.44	9.0	0.23	0.2	0.18	1.5
Float at sp. gr. 3.3	37.6	76.56	83.0	11.05	12.5	0.52	6.6
Float at sp. gr. 4.5	29.9	7.49	6.5	85.68	77.0	2.04	20.6
Sink at sp. gr. 4.5	7.4	7.28	1.5	46.63	10.3	28.56	71.3
Calc. head	100.0	34.69	100.0	33.28	100.0	2.96	100.0

This shows that 25% of the feed could be rejected for a loss of 9% fluorite, 0.2% barytes and 1.5% galena, and a fluorite concentrate assaying 76.6% CaF_2 at a recovery of 83% and a barytes concentrate assaying 86.5% at a recovery of 77.0% could be made.

The separator has had its most successful applications so far in separating heavy metals, eg copper, brass and lead from aluminium in feeds of shredded metal scrap. Applications in the separation of metal from slag are proceeding.

Magnetic Separation

One of the most important advances in magnetic separation has been the development of the family of high-intensity wet magnetic separators. The object of developing these separators was to enable minerals of small size and low magnetic susceptibility to be recovered economically. Until then the dry high-intensity separators could only recover these minerals down to a size of 200 mesh or a little beyond.

Jones in 1960[25] gave details of a small-scale magnetic separator in which fields calculated to be as high as 24,000 gauss were developed. The field was generated across a plate box filled with plates having serrated faces separated from one another by spacers whose size depended on the particle size and magnetic susceptibility of the feed. The first machines were semi-continuous but later, larger versions employed an annular plate box rotating past loading, washing and discharging stations set around the periphery to produce a continuously operating machine. Since that time many different versions have

Feed: (Itabirite with 53% hematite and 2.5% magnetite)

Grade	Capacity	Particle Size
39.7% Fe	1 Million TPY	−10 Mesh

Concentrate:

Grade	Recovery
67.9% Fe	95.5% (558,000 TPY)

Flow Sheet I:

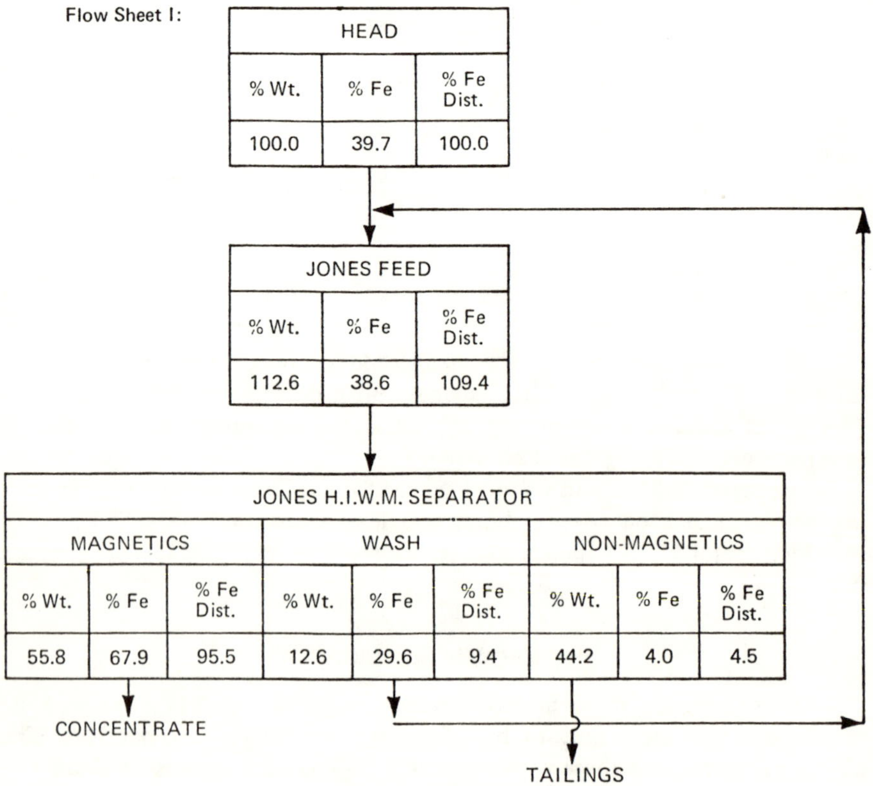

Fig. 14. Production data for Jones separator DP317

been produced by Gill in Australia, Jones in England (latterly by Humboldt-Wedag in Germany), by Carpco in the USA and by Krupp in Germany. In the main these differ only in the way in which the magnetics are trapped in the annulus, with the exception of the Gill separator in which the magnetics are trapped on the surface of a central serrated cone.

There are numerous applications for this type of separator such as the removal of magnetic impurities from glass sands, of non-magnetics from ilmenite concentrates, magnetite from asbestos fibres, iron minerals from clays, upgrading haematite ores, removal of haematite from phosphate ores before flotation, rejection of magnetics from a cassiterite concentrate and production of iron superconcentrates. Minerals such as scheelite, chromite, biotite, mica and columbite are also amenable to the technique.

Data relating to the performance of a Jones Humboldt-Wedag machine model DP317 treating 120 tonnes/h of itabirite iron ore assaying 39.7% Fe and producing 67.9% Fe concentrate at 95.5% recovery have been reported[26] and are shown below. The flowsheet for the operation is shown in Fig. 14.

Summarising the costings which are given in detail in the paper these are given in Table 19.

Table 19.

	$
Cost of separator	347 000
6 Centrifugal pumps delivered to German border only	24 000
Freight to Montreal (Customs duty excluded)	20 000
Total	391 000
Building & installation	95 000
Total capital investment	486 000
Depreciation & interest rate	53 168
Calculated from	

$$K = \frac{A}{100}\left[\frac{100}{n} + \frac{p(n+1)}{2n}\right]$$

where A = capital cost $486 000
p = annual interest rate 8%
n = service life 15 years.

	$ p.a.
Labour 1/5 man day 20% × $3 × 15 000	9 000
Power 1.8×10^6 kWh at $0.01/kWh	18 000
Water Total requirement 5 000 000 yd^3/year (3.82×10^6 m^3/year)	
With 90% recirculation new demand 500 000 yd^3/year (382 000 m^3/year) for make up at $0.02/$yd^3$	10 000
Cost of recirculation of water	7 000
Cost of grooved plates	20 000
Mechanical & Electrical maintenance	11 000
Total operating and maintenance	75 000
Total operating and capital costs p.a. =	$128 000

From the flowsheet Fig. 14 and the above data the costs per ton of feed and concentrate are given in Table 20. The costs are based on rates prevailing in Canada in 1972 and would have to be re-worked to take account of current labour, capital and interest rates.

Table 20: Treatment costs for iron ore

Cost Description	Iron Ore Concentration (Feed rate 1 million TPY) Feed $/Ton	Conc $/Ton
Equipment F.O.B. Montreal	0.043	0.080
Building and Installation	0.010	0.015
Capital Cost	0.053	0.095
Labour	0.009	0.016
Power	0.018	0.032
Water	0.017	0.030
Supplies	0.020	0.036
Maintenance	0.011	0.020
Operating Cost	0.075	0.134

Similar calculations can be made for producing 342,000 tons of iron superconcentrates assaying 69.1% Fe, <1% insolubles at a recovery of 90.6% from 400 000 tons/year of spiral concentrates containing 85% haematite, 6% magnetite and assaying 62.5% Fe. In this case the costs become:

	$ ton feed	$ ton conc
Capital cost	0.133	0.155
Operating cost	0.188	0.220
Total cost	0.321*	0.375*

* Based on double pass operation.

Costs are shown below for the recovery of values from 700 000 tons/year of spiral concentrator tailings containing approximately 6% goethite, 20% haematite, and 4% magnetite assaying 21.2% and producing 170 000 tons/yr of concentrate assaying 64.1% Fe at a recovery of 75.6%. The iron values are mainly −150 mesh in size.

	$ ton feed	$ ton conc
Capital cost	0.076	0.304
Operating cost	0.106	0.426
Total cost	0.182	0.730*

* Based on tailings already within the plant.

Data supplied by another manufacturer, Boxmag Rapid of Birmingham[27], England, for their range of separators gives a useful idea of how costs vary for capacities ranging from 6.5 tonnes/h for their SHWI single-pole model up to 104 tonnes/h for their SHWX8 8-pole machine. The data in Table 21 are based on the separation of Brazilian itabirite ore with feed rates of 56 940 tonnes/year for the SHWI model to 911 000 tonnes/year for the SHWX8 model. The magnetic products would amount to 28 470 tonnes and 455 525 tonnes/year respectively.

It should be borne in mind that the costs for the Jones machine include transport to Canada and operation within that country and assume an interest rate of only 8% pa and depreciation over 15 years while the Boxmag-Rapid figures refer to operation in Gt. Britain where the capital charges are based on interest at 15% and depreciation over 10 years. However, sufficient detail has been given for these corrections to be made. The separations quoted are, however, very easy for these machines to make at high rates and

Table 21.

	SHWI £	SHWX8 £
Capital costs		
Magnetic separator (power supply and controls)	17 250	100 000
Pumps	600	9 600
Buildings at £22.50/Ft^2	3 600	11 700
Installation and start up	900	2 925
Total	22 350	124 225
Depreciation and interest costs		
Annual depreciation over 10 years	2 235	12 423
Annual interest at 15% p.a.	1 845	10 166
Total	4 080	22 589
Operating costs p.a.		
Labour man hrs/day	2	4
Costs at £12 man/day	365	730
Electrical power cost at £0.017/kWh	1 485	15 145
Spares 5% of magnet+pump costs	863	5 000
Maintenance 2½% of total capital cost	432	2 500
Water requirement 000's gal/yr	47 300	756 900
New water assuming 50% recirc and water at £0.28/1000 gal	6 811	108 960
Total operating costs	9 956	132 335

Cost	SHWI Cost £ tonne feed	SHWI Cost £ tonne conc	SHWX8 Cost £ tonne feed	SHWX8 Cost £ tonne conc
Capital	0.07	0.14	0.02	0.04
Operating	0.18	0.36	0.15	0.30
Total	0.25	0.50	0.17	0.34

other materials will prove to be more difficult, thus throughputs will be lower and costs higher. Boxmag-Rapid have produced an interesting table which shows the approximate treatment rates per pole for the SHWX series of separators which have 4, 6 or 8 poles (Table 22). The magnetic susceptibility,

Table 22: Approximate treatment rates per pole for various materials

Material	K gauss	tonnes/hr	m^3/hr	Imp gal/min
Kaolin*	19–24	3.0	10.8	40
Feldspar*	16–18	5.5	16.2	60
Haematite USA	16–18	6.5	16.2	60
Ilmenite	13–15	9.0	21.6	80
Columbite	13–15	10.0	21.6	80
Wolframite	13–15	11.0	21.6	80
Chromite	11–12	12.0	27.0	100
Itabirite	9–10	13.0	32.4	120

* In applications where the quality of the magnetic products is not important and the quantity removed is small the figures can be doubled.

sizing and specific gravity of the material will all have an effect on throughput. The figures given are not a guarantee of performance, rather a guide as to what might be expected under average conditions. Proper testwork with representative samples is necessary to obtain specific costs for any particular problem.

Flotation

Advances in flotation arise along several fronts, theoretical advances, reagent developments which in turn lead to improved treatment flowsheets, equipment design and in process control of flotation. The theoretical advances will not be dealt with here but it should be realised that advances in this field can, in the long term, have far reaching effects in the other fields mentioned above.

The tendency in flotation machine design, as with grinding mill design, has been to construct larger machines. For many applications these are of conventional design, but occasionally the design has had to be modified to meet special needs.

The techniques employed in mineral flotation have, with modification over recent years, been rapidly extended into realms such as the flotation of oils, ions, precipitates and very fine particles. Many of these demand different operating conditions from those found in the normal ore flotation cell, ie finer bubble sizes, more quiescent conditions, deeper cells etc.

In the field of flotation cell design, the mid-sixties produced a flurry of different designs of flotation impellers such as the Wemco 1 + 1, Denver DR, Minemet cross barred rotor, the Boliden FR rotor etc. At this time the largest cell sizes were about 100 ft^3 (2.8 m^3). At the present time cells of conventional design up to 600 ft^3 (17 m^3) are being used mostly in order to treat very high tonnages per day of very low-grade copper ores. For example, at the Gibraltar, BC, concentrator, 36 000 tonnes/day are floated in sixteen 600 ft^3 (17 m^3) Denver cells, scavenged in sixteen 300 ft^3 (8.5 m^3) cells and cleaned and re-cleaned in sixteen 100 ft^3 (2.8 m^3) cells. The justification for using these large cells is to reduce the total number of units with their ancillary motors, switches, power lines, walkways etc and effect savings in space and also in higher efficiency in terms of hp per ton of throughput.

At Bougainville[5], which treated 78 000 tonnes per day in 1972, roughing is carried out in one hundred and eight 600 ft^3 (17 m^3) cells laid out in six banks of 18, scavenging in fifty-two 100 ft^3 (2.8 m^3) cells in four banks of 13; cleaning is in thirty-nine 100 ft^3 (2.8 m^3) cells and re-cleaning in a further twenty-six 100 ft^3 (2.8 m^3) cells. Walkways are provided between each bank of cleaner cells and the rougher cells are laid out lip to lip with a common concentrate launder. By adopting the largest size of cells for roughing and the layout above a 25% saving in floor area is estimated to have been made over a conventional layout.

At Opemiska expansion of milling facilities was made to increase capacity from 2000 to 3000 tonnes per day floating copper ore at a grade of 2.25% Cu with a grind of 60% −200 mesh. Maxwell cells are used throughout with 2000 ft^3 (56.6 m^3) cells for roughing, 700 ft^3 (20 m^3) for scavenging and 300 ft^3 (8.5 m^3) and 170 ft^3 (4.8 m^3) for cleaning. Previous power consumption was 400 hp for 2000 tonnes per day and present consumption is now 280 hp for cells plus air supply for a throughput of 3400 tonnes per day. Total maintenance for the flotation circuit was $1,100 per year over the last 3 years. Floor space required for the cells and walkways is 2000 ft^3 (186 m^3). Maxwell cells are also used at Mainbridge.

At Nacimiento[28] a major change in flotation cell size was made with the invention of the Maxwell cell which takes the form of a cylinder measuring 14 ft dia × 14 ft high (425 × 425 cm) with a volume of 2000 ft^3 (56.6 m^3). This was used to expand the existing cell capacity of 2400 ft^3 (68 m^3) in order to recover sulphide copper by the addition of two of these giant cells. This effectively doubled the resident time in the circuit for only 400 ft^2 (37 m^2) of extra floor space and the machines were accommodated within the existing flotation plant. For an outlay of $100,000 added to a $5 million mill the capacity was increased by one third.

At a Canadian iron ore concentrator, 20 Wemco No 144 cells of 500 ft^3 (14 m^3) treat 10,000 tonnes per day of feed for a specific power consumption of 0.57 kWh/tonne and at Silver Bell, Arizona, 144 No 66 size cells were replaced by 32 cells of 120 size handling 15,000 tonnes per day; the saving was a reduction in total hp from 1,150 to 736, ie a daily power saving of 7,400 kWh.

In Australia, an injection flotation cell, the Davcra[29] cell has received some attention. In this cell, which takes the form of a tall rectangular tank,

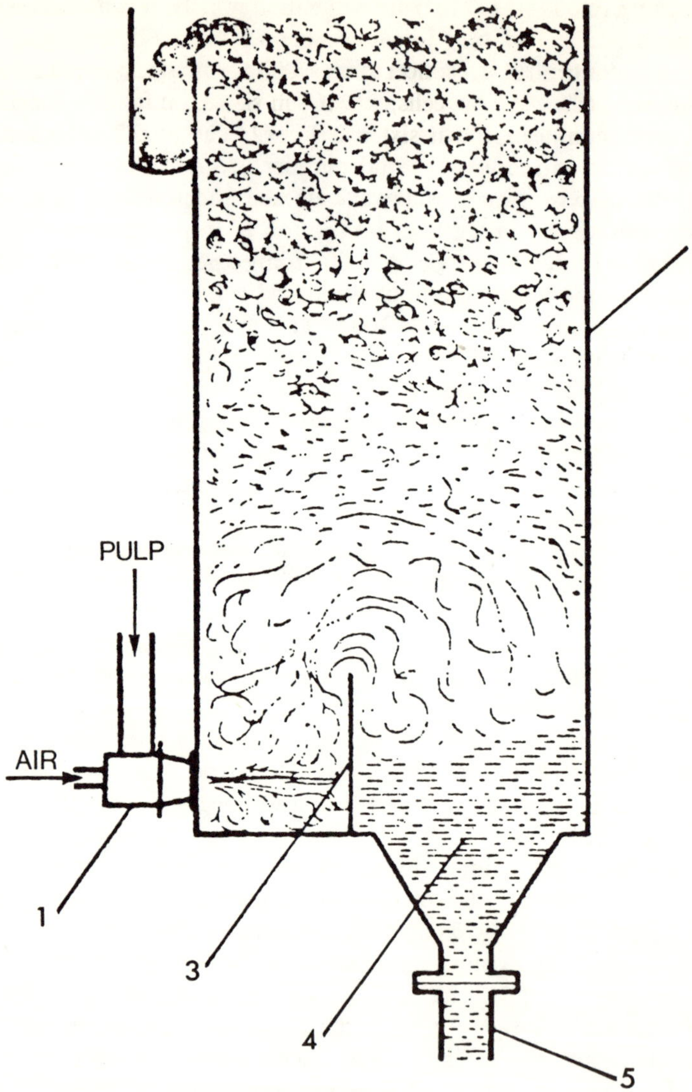

Fig. 15. Sectioned view of DAVCRA flotation cell—1. Nozzle body, 2. Tank, 3. Baffle, 4. Deaeration zone, 5. Tailing discharge.

the pulp and air are introduced through an injector at the base of the cell, the cross section of the cell is shown in Fig. 15. It is claimed that the air/pulp mixing which results from this arrangement is more efficient than in the conventional machine.

In 1967 a fully integrated lead flotation circuit comprising five roughers and two cleaner cells was installed at the Zinc Corporation property, Broken

Hill. This took one third of the whole feed to the plant (60 tonnes per hour) and produces acceptable grade concentrate with tailings grades close to those achieved in conventional cells. Cells with capacities of 200 tonnes per hour have been installed at the Zinc Corporation's concentrator with great success but their use has not spread to other countries. It is understood that there are some wear problems associated with the baffle plate.

A further new design of flotation cell, the Nagahm cell, has been developed in Japan to deal with the recovery of ions, precipitates and ultrafine particles arising in overflow from zinc concentrate thickness at the Kamioka concentrator of Mitsui Mining and Smelting[30]. This was done in order to overcome a pollution problem and to try to recover values which were being lost. The effluent contained cuprocyanide complex, cyanide ions and fine particles of ZnS and CdS together with ethyl xanthate and frother.

The requirements of a cell to recover this type of material are that a large volume of bubbles must be generated, and the bubbles must be small and evenly distributed, the cell must be non-turbulent and must be capable of discharging large volumes of froth. The performance of the Nagahm cell with respect to aeration compared with some others is shown in Table 23.

Table 23: Comparisons of Aeration Using Different Flotation Machines

Type and Dimension of Flotation Machines	Fahrenwald No. 24	Agitair No. 48	Warman No. 48	V-flow No. 48	Nagahm 66 (Water Level) 760 mm	650 mm
Impeller Diameter (mm)	610	685	430	440	390	
Rotation of impeller (rpm)	280	157	400	285	645	
Peripheral Speed of impeller (m/min.)	536	348	540	390	790	
Aeration Rate ($l/min.m^2$)	320	900*	744	1,095	1,785	2,115
Power consumed (kW/h)	4.1	5.84	3.4	2.96	6.67	6.63
Power per m^3 of aeration ($\times 10^2 kW/m^3$)	152	72.7	53.7	30.3	28.6	23.7

* The aeration rates of Agitair and Warman flotation machines are the most suitable ones, under the conditions at which the grades of flotation tailings are lowest.

The economics of the process are shown in Table 24. Note that ipu means ions, precipitate and ultrafines. The capital cost of the installation was 20.9 million yen, about $70,000 and second-hand equipment was used wherever possible. The saving per annum is of the order of $50,000. For a full description of the process Nagahama's paper should be consulted. This example illustrates the way in which flotation is becoming closely linked to pollution problems and it is probably in this area in which further expansion will take place in the future.

Other methods of floating fine particles by generating small bubbles are appearing. Glembotsky *et al*[31] describe a process where electrolytic action

Table 24: Economics of IPU flotation treatment

Income (yen/year)		Expenditure (yen/year)	
Copper Recovered from Effluent	2,847,000*	Reagents	2,724,000
Saving of NaCN	1,716,000	Power consumption	2,687,000
Zinc recovered from Effluent	9,608,000	Others	936,000
Increase of Zn Conc. in Differential Pb-Zn Flotation	8,427,000		
Total	22,598,000		6,347,000

Profit: 22,598,000 − 6,347,000 = 16,251,000 (yen/year)

* Copper is recovered as 1st-stage purification residue of zinc electrolytic process, with the estimated recovery of 50%.

is used to generate small bubbles to float ultrafine particles, no costs are given but increased recovery, quality of concentrate and higher flotation rates are said to outweigh power costs for generating the bubbles.

In the application of flotation processes, probably one of the most significant advances in the last few years has been the successful application of flotation techniques for the recovery of fine cassiterite. A paper presented to the Tenth International Mineral Processing Congress by Moncrieff *et al*[32] outlined the techniques used at the Janes Mine Cornwall, Renison Bell Tasmania and Union Tin and Rooiberg in S. Africa. The flowsheets are quite complex and the original paper should be consulted for details. This process is only suitable for very fine cassiterite which is difficult to recover in normal gravity circuits and does not supplant the normal methods for recovery of coarse cassiterite.

A summary of the costs is given in Table 25 which shows the direct costs of operating the flotation circuits in terms of the local currencies.

The installed horsepower at the four operations is summarized below. These figures exclude standby pumps, tailings reclamation (South Africa), grinding (Wheal Jane) and tailings disposal. The figures for flotation include sulphide scavenging and concentrate handling and upgrading (desliming and magnetic separation at Wheal Jane).

	Feed to desliming, ton/h	Installed horsepower Desliming	Flotation, etc
Renison	25.5	195	561
Union Tin	8.9	170	126
Rooiberg	8.2	183	97
Wheal Jane	20.0	100	610

It has been necessary to restrict the scope of the paper to a few of the developments which can have a significant economic impact on processing

Table 25: Direct operating costs from Gold Fields group mines using tin flotation

	Renison	Union Tin Emulsified	Rooiberg	Wheal Jane
Collector	SM 119	PTAA	PTAA	Alcopol R.540¶
kg/ton††	0.42	0.25	0.5	0.66
Price/kg	$2.53	R4.00	R4.00	£0.36
Costs per ton of flotation feed	$ A	Rand	Rand	£
Reagents				
Collector*	1.06	1.0	2.0	0.24
Others	0.23	0.3	0.6	0.14
Total	1.29	1.3	2.6	0.38
Other	0.28†	1.0‡	1.2‡	0.4
Total operating**	1.57	2.3	3.8	0.78
Maintenance	0.7	§	§	§
Total	2.27			

* Excludes royalties where applicable. † Excludes power. ‡ Includes desliming. § Insufficient data since start-up. ¶ AP 845 consumption was about 0.9 kg per ton and its price about £0.48 per kg. ** Excludes any allocation of mill overheads. †† Renison and Wheal Jane collector consumption figures refer to the reagent as received, but Union Tin and Rooiberg figures refer to PTAA content of the reagent. SM 119 as received contains 25% water; PTAA as received contains 45% PTAA; R.540 and AP 845 as received contain 35% R.540 or AP 845.

costs and the omission of a field does not necessarily mean that no advances have been made. An idea of the trends in each subject have been given together with specific examples of practice and, where available, costs and performance figures given. This latter area is the one in which detailed information is difficult to find and it is almost impossible to discover how two different machines or processes will work on the same ore so that direct comparisons can be made.

For the sake of brevity much has had to be omitted regarding the finer points of the operations discussed and the appropriate references should be consulted for this.

Discussion Summary

There was during the discussion mention of autogenous grinding problems experienced in Turkey, Mexico and Canada, chiefly the inability to deliver designed performance. The failures were attributed to change of ore type from that tested in pilot plants, to over-optimistic specifications, to flowsheet design faults, to the lack of automatic control to deal with rapid changes in ore types in the feed and to problems arising from the need for skilled operation and maintenance. Discussion also centred on operational problems incidental to dry operations where excessive dust clogs filters and fine sizes cannot be efficiently concentrated.

REFERENCES

1. Burt R O. Fine sizing of minerals. Mining Mag. *128* (6) June 1973, p463–465.
2. Hepworth W E. Design & performance of a special multideck screen. Quarry Management and Products. Feb 1974 p39–42.
3. British patent 124904 1970.
4. Rushton M W and Mackay K E. Nchanga Mine Northern Rodesia. Trans. Inst. Min and Met. 70(3) 1960 p77–100.
5. Anon. Giant machines in mill. Wld. Mining *25* (11) Oct 1972 p65–71.
6. Shoemaker R S. Minerals Processing in 1973. Min. Congr J. *60*(2) Feb 1974 p24–29.
7. Anon. Island Copper project. Min. Mag. *127* (4) Oct 1972 p344–349.
8. Nice R W. Concentrator start up at Similkameen Bull.Canad. Inst Min & Met. *67*, 749, Sept 1974 55–61.
9. Fish R. Canadian Mining Journal Reference Manual 1974. National Business Publications Ltd, Montreal 1974.
10. Anon. Bethlehem Copper, Canada's first major open pit porphyry copper. Min. Mag. *129* (5) Nov. 1973 p406–410.
11. Falström P H. Lead and zinc flotation practice at Boliden in the mining and concentrating of lead and zinc. New York AIME 1970 p668–694.
12. McDermott W F et al. The dollars and sense of autogenous grinding. Min. Engg. *24* (11) Nov 1972 p46–50.
13. Øyasaeter O. Pebble grinding practice I and II. Wld Mining *25*(8) July 1972 p32–39 and 25(9) Aug 1972 p66–69.
14. Anon. Erwo milling limits overgrind—uses less power. Wld Mining *27*(11) Oct 1974 p66–67.
15. Private communication Waagner-Biro, Vienna.
16. Amsden M P, Pickard F G, Deluca F R and Molloy P. Panel on rubber mill linings. Canad. Min. J. *6*, June 1974 p42–45.
17. Yakolef V P et al. Tovetnya Metallurgia 1974 No 2 p21–22.
18. Balint A. The beneficiation of minerals with electronic separators in Proc. First Industrial Minerals International Congress, London 1974. Metal Bulletin Ltd p53–65.
19. Hintikka O V and Balint A. Optical separation of limestone in Southern Finland by Lohjan Kalkkitehdas Oy. Trans. Soc. Min Engrs *250*(3) Sept 1971 p203–207.
20. Private communication Hunt and Moscrop, Manchester.
21. Chaston I R M. Heavy media cyclone plant design and practice for diamond recovery in Africa. In Proc. 10th Int. Min and Met Congress: London Inst of Min. & Met. 1975 p257–276.
22. Burt R O and Ottley D J. Fine gravity concentration using the Bartles-Mozley concentrator. Int. J. Miner. Processing 1, 1974 p347–366.
23. Muller L D and Sayles C P. Processing dry granular materials. Min. Engg. *23* (3) Mar. 1971 p54–56.
24. Douglas E and Walsh T. New type of dry heavy medium separator. Trans. Inst. Min and Met *75* Sept 1966 C226–232.
25. Jones G H. Wet magnetic separator for feebly magnetic minerals in Proc. 5th Int Min Proc. Congr. London. IMM 1960 p317-372.
26. Bartnik J A and Stone W J D. Capital and operating costs of the Jones high intensity wet magnetic separator plants for iron ore concentration. AIME Annual Meeting Duluth 1972.
27. Private communication Boxmag Rapid Ltd, Birmingham.
28. Ealy G K. Concentration of copper oxides by flotation at Nacimiento. Min. Congr. J. *59*(3) Mar 1973 p63–66.
29. Gauci G. A comparison of ore dressing practices at Broken Hill Australia in Mining and Concentrating of Lead and Zinc. New York AIME 1970 p373–400.
30. Nagahama T. Treatment of effluent from the Kamioka concentrator. Bull. Canad. Inst. Min and Met. *67* 744 April 1974 p79–89.
31. Glembotsky T et al. Selective separation of fine mineral slimes using the method of electro flotation. In Proc. XI Int. Min. Proc. Congr. Cagliari 1974 Paper 20.
32. Moncrieff A G et al. Development and operation of cassiterite flotation at mines of the Consolidated Gold Fields Group in Proc. 10th Int. Min. Proc. Congr. London. Inst. Min and Met 1975 p565–592.

PAPER 7

RECENT ADVANCES IN THE FIELDS OF IRON ORE PROCESSING AND SPONGE IRON PRODUCTION

by B. R. Nijhawan

AN IRON ORE IS a natural mineral aggregate from which iron can be extracted economically. The economics depend largely on cost saving and productivity increases brought about by beneficiation and preparation techniques. These involve processing and blending ores, making best use of the most favourably available raw materials to yield blast furnace burdens (smelting charges) characterised by uniformity, generally high iron content and favourable physical and chemical properties.

The iron ores of commerce contain in general between 55 and 65% iron with the important exception of the minette ores which run about 28% iron. Some low-grade ores can be upgraded or concentrated by mineral processing methods and will be economic if situated near to markets and favoured with adequate infrastructure. The minette ores cannot be upgraded but are blended for use.

The processing methods usually yield concentrates of iron minerals of more or less fine grain size which have to be prepared for use as furnace burden by various processes of agglomeration (pelletising, sintering, briquetting).

The advent of high grade concentrates (high iron and low gangue contents) and the availability of some equally pure high grade lump ore coupled with the increasing scarcity or local unavailability of coking coal has led to the successful development of direct reduction processes using natural gas. These have the advantage of being economic at moderate scale. The product (sponge iron or metallised pellets) can be fed to electric arc furnaces, by passing blast furnace technology and leading to the viability of "mini mills".

Iron Ore Processing

Beneficiation

The major constituents of iron ores, apart from iron oxides and other iron compounds are silica, alumina, lime and magnesia. Their presence affects the technology and cost of smelting. Other impurities naturally occurring include phosphorus, sulphur, manganese, titanium and other metallic elements. Removal of some of these constituents or adjustment of their proportions can be of considerable economic value.

This beneficiation, or improvement of grade and amenability (and therefore of value) of iron ores, can be achieved by a variety of mineral processing (concentration) methods with or without additional metallurgical methods such as pelletising, sintering, and briquetting and the production of metallised pellets.

Beneficiation can be effected in some cases by simple washing and screening processes to remove clay, debris and low grade fines. Other methods are described in the following sections.

Concentration

Iron ores can be upgraded if the richer iron bearing minerals (the oxides) can be liberated by crushing and grinding and then separated from the accompanying unwanted minerals (gangue) and impurities by known mineral dressing procedures based on differences in specific gravity, magnetic and electrical properties and surface characteristics. These procedures include the use of heavy medium separation in cones, drums or cyclones, of jigs and spirals, of magnetic and electrostatic machines and of flotation.

Use of such processes has resulted in the development of significant production from the low-grade Lake Superior type of ore (taconite, itabirite, jaspilite). Not all of these deposits respond to the same concentration process flowsheets. Some remain intractable. Only recently has a combined selective flocculation and flotation process been found feasible for one of the martite/hematite deposits of the Lake Superior region.

Pelletising

Iron ore fines are not acceptable as a charge or part of a charge to blast furnaces or for some of the direct reduction processes but can be rendered usable by processes of agglomeration. At the mine, where fines are produced by screening of the crushed ore or as concentrates from an upgrading process, the method adopted is pelletising.

The fines are mixed with binding agents and water and fed to rotating disc or drum pelletisers where the mixture coalesces and is rolled up into spherical pellets. They are then indurated (hardened) by either hot or cold means.

In the hot process the binders are usually bentonite and lime and the heat hardening process involves drying, pre-heating and final heating to a temperature of 1,250–1,350°C. The heating equipment can be of the shaft furnace, travelling grate or circular grate type. The last is the latest development and is the least costly and the most compact and simple.

Cold induration processes include the Swedish Grancold process which has found acceptance due to its low capital cost and its applicability to local fabrication and operation under licence in developing countries like Brazil and India. The binder is up to 10% of Portland, Pozzualanic or slag cement.

The "green" pellets are coated with iron ore concentrate to prevent sticking and clustering and allowed to cure and harden for up to a month.

Other processes include the carbonate bond process (admixture with lime and hardening at 120–150°C in a carbon dioxide atmosphere under pressure), the corrosion bonding process (addition of iron chips and sodium chloride) and autoclave bonding with tar and pitch. These processes and the methods of evaluating pellet properties are receiving considerable research and development attention

Development is also proceeding on processes of pre-reduction or metallising of pellets to very high iron contents to yield an enriched feed to blast furnace and electric iron smelting furnace plants and for use in steel making as an alternative to scrap. Pelletising was introduced in 1944. By 1964, production was 45 million tons a year and in 1974 production was 172 million tons, of which 3.35 million tons were pre-reduced. During 1974, eight new pelletising plants came on stream, of which two were in a developing country (Mexico).

Pellets and pre-reduced pellets can be transported over great distances in the same way as iron ores. They represent a valuable commodity in iron ore trade and a means of adding value to iron ore production.

The cost of pelletising varies widely with local conditions and size of plant. Economies of scale are to be expected but recent price escalations and general inflation have increased operating costs to more than double those obtaining in the sixties. Typical costs are given in Table 1.

Table 1: Cost of Pelletising (*US$*)

Cost elements	Unit	Rate	Consumption	Cost/ton pellets	
1. Iron ore fines cif.	ton	18	1.14	20.52	
2. Bentonite	,,	41	0.012	0.492	
Total materials—					21.012
3. Labour and supervision				0.30	0.30
4. Services					
i) Natural gas	Nm^3	0.0125	40	0.50	
ii) Power	kwh	0.0206	35	0.72	
iii) Others				0.28	
Total services—					1.50
5. Operating supplies				0.26	
6. Repair and maintenance				0.50	0.76
Total cost above material					2.56
7. Overheads				1.2	1.2
Total cost of production excluding capital depreciation and interests					24,722

Sintering

Sintering is a process used within a smelter complex for agglomerating iron ore fines and smelter dusts and fine residues which are produced during crushing and screening of raw ore or at other points in the process line. The fines are caked into a porous but strongly coherent structure by controlled heating. A mixture of the fines with coke breeze is fed on to a moving grate and ignited by means of a gas flame. Combustion proceeds by controlled down-draught through the grate. Admixtures of limestone or dolomite may be used to produce self-fluxing sinter. The cost of sintering amounts to around US$4 per ton of sinter but this is normally part of the smelting cost and would not be considered in connection with ore production and processing.

Briquetting

This is a third method of agglomeration used sometimes in steelworks to produce a dense material suitable for charging into open-hearth furnaces as a substitute for raw ore. The briquettes are produced by compression of fines with or without binders and indurated at moderate or sintering temperatures respectively.

Iron Ore Preparation

The improvement of productivity in the production of iron in blast-furnaces in the last two decades has been influenced very largely by preparation of the feed. Iron ores are no longer indiscriminately charged but are carefully selected, crushed and graded for size and removal of fines, stock-piled, blended and re-claimed in a manner such as to provide as rich as possible a feed in a uniform, most economically and rapidly reducible state. Sintering of fines and the admixture of sinter into the burden contribute largely to this improvement of efficiency but these processes of preparation relate to the operation of iron and steel plants. The main contribution at the mining end is the recovery of high grade concentrates from lean ores and the production of pellets.

Direct Reduction and Sponge Production

Conventional reduction of iron ores is performed in blast furnaces at a very high temperature. The gangue minerals are separated as a liquid slag and the iron is recovered in the molten form. On the other hand, direct reduction is carried out in the solid state, without melting, at considerably lower temperatures.

Broadly, there are two types of direct reduction. The first is gaseous reduction of very pure ores. Natural gas is used and the product is sponge iron of high purity. Only pure ores can be used, since the process does not

Table 2: Cost of Production of Sponge Iron Gaseous Direct Reduction Plant at 100% Capacity Utilisation in US Dollars

Cost element	Unit	Rate	Specific Consumption	Cost per ton of sponge iron	
1. Pellet	ton	24.77	1.4	34.678	
2. Natural gas	Nm^3	0.0125	460	5.750	
Total materials—					40.428
Cost above Materials					
1. Labour and supervision					
Services				0.450	
2. Power	kwh	0.0206	30	0.618	
—others excl. transport				0.30	
Total services—				1.368	
3. Operating supplies				0.4	
4. Repair and Maintenance				0.75	
Total cost above Material—					2.518
Overheads				0.60	0.60
Total cost of production excl. capital depreciation and overheads					44.914 or say 45 US$ per ton sponge

provide a means for separation of gangue. The original process of this type is the HyL/lc which has been followed by Midrex, Purofer, FIOR and HIB based on the use of natural gas; other processes based on heavy oils or gasification of solid fuels are under active development. In some cases the feed is provided in fluidised form or as briquettes. The second type utilises solid reductant in rotary kilns, typified by the Krupp and SL/RN processes which can be used for such ores as titaniferous iron ores or iron sands or for low grade siliceous ores, steel mill sludges and wastes, etc.

A distinction has to be drawn between two types of product. One is a highly metallised sponge, pellet or briquette which can be used directly for steelmaking in the electric arc steel furnace. Such material is subject to ready re-oxidation and therefore cannot be transported any considerable distance unless protective atmosphere and/or sponge passivation treatments are employed. The other type is a partly metallised product which can be the feed for electric submerged arc iron smelting furnaces.

Where the product is a partly metallised material forwarded to a further reduction process, the preparatory stage is referred to as pre-reduction as in the ferro alloy industry or in the treatment of nickeliferous laterites.

In the case of the iron blast furnace, considerable pre-reduction takes place in the shaft of the blast furnace itself and so the advantages of a pre-reduced charge are not so marked as in the case of electric submerged arc furnace for iron-smelting where the shaft height and effects are minor. The charge of a fully and directly reduced sponge to the blast furnace is another question which needs a separate techno-economic study and appraisal.

The attraction of direct reduction processes lies in the possibilities for the use of under-used or wasted natural gas resources and flare gas and also for the setting up of small steel making units independent of the supply of scarce coking coal.

The processes therefore have great attraction for developing countries. The foremost (HyL) originated in Mexico in the nineteen fifties. That and the variants have been installed in Argentina, Brazil, Peru and Venezuela and also in N. America, S. Africa, Japan, Europe and USSR. (NB Historically the Krupp Renn process preceded these, having been developed in the nineteen thirties but found only limited application in the circumstances of its time; currently, it is almost entirely out of use.)

Direct reduction processes have recently come into prominence in developing countries where oil, natural gas and flare gas are available. Installations are on order for Iran and Iraq and under consideration for Abu Dhabi, Kuwait, Qatar, Saudi Arabia, Turkey, Algeria, Egypt, Libya, Nigeria and Tunisia.

The following factors are important:

(a) Direct reduction processes and their contributions to the growth of the world's steel industry should be viewed in proper perspective. These processes do not offer a universal panacea for all ills retarding the growth of the steel industry, particularly in developing countries.

b) Some of the direct reduction processes are still cutting their teeth and remain techno-economically unproven despite contrary claims particularly those based on solid reductants (non-coking coals and fossil fuels) using rotary kiln installations. Utmost caution is required in advocating their applications, *inter alia*, in developing countries some of which have already paid a heavy price in plant installations scrapped within a year or two of their commissioning.

c) Direct reduction *per se* should be considered on the basis of offering possibility of *relatively minor complementary growth* to the conventional iron and steelmaking processes; the latter undoubtedly provide the main sinews of the steel industry's current status and future growth and will continue to do so in the decades to come.

d) Direct reduction processes will in the foreseeable future not be able to displace blast furnace technology and any such talk on technical platforms tends to mislead many in the developing countries.

e) Direct reduction processes should be considered for supply of sponge for steelmaking to feed the merchant mills in view of the limitations of the module sizes inherent in direct reduction processes (much lower in solid reductant based processes than for gas based direct reduction processes).

f) The choice of technology (direct reduction processes and/or conventional iron and steelmaking) should be related to the scale of operations of the steel industry and the markets it seeks to serve (home and export markets).

An appraisal of some solid reductant based processes for the production of sponge iron was made by UNIDO Secretariat* in 1973; this paper discussed the reasons for the failure of the following solid reductant based rotary kiln plants and their closing down soon after their start up periods with heavy financial losses and in the case of Inchon the loss of many lives as well:

a) Skopje Iron and Steel Works (Yugoslavia);
b) Inchon Iron and Steel Company (S. Korea);
c) Falconbridge Nickel Company (Canada).

Economic Aspects of Direct Reduction

There are no universal yardsticks to assess the economics and production costs of direct reduction including capital and operational costs; these will again differ in the case of gaseous direct reduction processes, vis-à-vis solid reductant based rotary kiln direct reduction processes; the following Table No. 2 depicts the general indications of the cost data for a gaseous direct reduction plant.

Some countries may not have adequate natural gas resources but may have surplus naphtha which can be used for sponge production even at the prevalent *high* naphtha prices. The following Tables 3 and 4 depict the *economics* of sponge production based on gasified naphtha; a sponge plant is currently under installation in Zambia based on naphtha as the reductant.

It would not be out of place to refer to the most leading direct reduction processes and their capacities (in operation and under construction) as of 1975.

Process	Direct reduction and electric furnace plants	DR plant operational	Units under construction	Rated capacity tpy
HyL	10	6	7	8,000,000
Midrex	8	5	5	6,500,000
Purofer	2	—	2	700,000
Armco	2	1	—	330,000

Conclusion

In concluding this paper, it is stated that the approach of UNIDO to new technological developments, such as direct reduction is one of enlightened

* An appraisal of some of the direct reduction processes for the production of sponge iron—ID/WG.146/118—UNIDO Secretariat (September 1973) Paper for UNIDO's Third Interregional Symposium on the Iron and Steel Industry, Brazil, October 1973.

Table 3: Estimated Operating Costs to Reflect Current Light Naphtha prices*

165,000 Annual Metric Tons Fe in Sponge Iron Metallisation—85%

	Unit Cost U.S. $	Unit Consumption Metric Ton Iron	Annual Cost U.S. $	Cost U.S. $ Metric Ton Iron
1. Lump Ore, sized +1/2″–2″ 280,000 Annual Metric Tons 62% Fe	$12.0/Metric Ton cif. ($0.193/Metric Ton Iron Unit)	1.7 Tons	$ 3,336,000	$20.2
2. HyL Reduction				
Light Naphtha	$11.0/BBL	5.0 BBLS	9,075,000	55,000
Electric Power	0.01/kwh	20 kwh	33,000	0.200
Water	0.10/1000 Gals.	2,140 Gals	36,300	0.220
Chemicals, Catalysts & Misc.			41,250	0.250
Maintenance Materials & Labour			412,500	2.500
Operating Labour	0.625/Manhour	108,000 Manhours	67,500	0.409
Operating Supervision	1.25/Manhour	11,000 Manhours	13,750	0.083
General Overhead 100% Labour & supervision			81,250	0.492
Total HyL Reduction			$9,800,550	$57,154
3. Grand Total			$13,136.550	$77,354
Not Included—Depreciation, Taxes & Insurance, Royalty, Cost of Sales and Overhead				
4. Estimated Annual Consumptions				
Lump Iron Ore (sized +1/2″–2″) (Includes a 5% fines loss allowance)		280,000 Metric Tons		
Light Naphtha		825,000 BBLS		
Electric Power		3,300,000 kwh		
Water		353×10^6 U.S. Gals.		
5. BBL = Barrel = 42 Gallons				
Gals = Gallons				

* Compiled from the technical literature by the author.

*Table 4: Estimated Operating Costs**

198,000 Annual Metric Tons Fe in Reduced Pellets
Metallisation—85%

	Unit Cost U.S. $	Unit Consumption Metric Ton Iron	Annual Cost U.S. $	Cost U.S. $ Metric Ton Iron
1. Iron Oxide Pellets, Sized +3/8"–5/8"	$16.0/Metric Ton	1.54 Tons	$5,040,000	$25.400
315,000 Annual Metric Tons 66% Fe	($0.242/Metric Ton Iron Unit)			
2. HyL Reduction				
Light Naphtha	$11.0/BBL	4.2 BBLS	9,147,600	46,200
Electric Power	0.01/kwh	16.8 kwh	33,200	0.168
Water	0.10/1000 Gals.	1.790 Gals.	35,400	0.179
Chemicals, Catalysts & Misc.			41,500	0.210
Maintenance—Materials & Labour			415,800	2.100
Operating Labour	0.625/Manhour	108,000 Manhours	67,500	0.341
Operating Supervision	1.25/Manhour	11,000 Manhours	13,750	0.069
General Overhead 100% Labour & Supervision			81,250	0.410
Total HyL Reduction..........			$9,836,000	$49,677
3. Grand Total..........			$14,876,000	$75,077
Not Included—Depreciation, Taxes & Insurance, Royalty, Cost of Sales and Overhead.				
4. Estimated Annual Consumptions				
Oxide Pellets (sized +1/2" = 2")		315,000 Metric Tons		
(Includes a 5% fines loss allowance)				
Light Naphtha		832,000 BBLS		
Electric Power		3,326,000 kwh		
Water		354×10^6 U.S. Gals		
5. BBL = Barrel = 42 Gallons				
Gals = Gallons				

* Compiled from the technical literature by the author.

technical and practical interest particularly in relation to their possible applications in developing countries.

It must be appreciated that developing countries with their meagre resources financial and technological, do have a valid case when they feel and rightly so that they should not be used as the base for commercialising technically and industrially unproven processes and technologies. UNIDO in its dedication to assist the developing countries endeavours to meet these requirements and build up technical linkages between the advanced and the developing countries in the fields of metallurgical industries including iron and steel industry and direct reduction sponge technology. A word of caution. Whilst the applications of the latest technological innovations, automation and computerised operations in developing countries are rightful ambitions and objectives, nonetheless the applications of fully *appropriate* technology should be encouraged depending upon the conditions and environments in each case and country. To illustrate this subject, it is recommended that the use of charcoal for iron smelting in relatively small blast furnaces (less than 200 ton per day capacity) should be fully supported in developing countries which have a good forest wealth and forestation programme. In a fast developing country like Brazil, for instance, some three million tons of pig iron is smelted in small blast furnaces using charcoal as the reductant and for heat input. In Western Australia, at Wundowie, an iron smelting blast furnace complex using charcoal has been operating profitably for the last two decades. In another fast developing country, India, in the Mysore Iron and Steel Works at Bhadravati, charcoal blast furnace operations have advantageously been carried out for years. In Malaysia, at the plant of Malayawata Steel, iron smelting has been successfully in operation for the last several years using charcoal made from rubber wood.

Discussion Summary

In discussion mention was made of the economic benefits of international iron-ore trade in connection with local smelting or direct reduction. Indonesia for example has iron-sands and natural gas but may find it advantageous to export the iron sands, which are difficult to treat, and import Australian pellets as feed for local direct reduction plant; South Korea exports high-silica ores and imports lump ore from India and Australia; New Zealand and the Philippines, however, are developing iron-sand feeds and solid reductant methods. Titanium-bearing iron sands may be smelted electrically to yield iron and high-TiO_2 slag. Briquettes compare unfavourably with sinter and pellets in terms of economy and utility. The Highveld (S. Africa) process for recovery of vanadium may also be of possible application for titaniferous iron sands containing vanadium. The development of Bolivian (Mutun) iron ore and natural gas supplies should be phased through stages beginning with beneficiation and export, to be followed by pelletisation and sponge production, and eventually by local steel production. The Strategic-Udy process has not been developed to the point of technical and economic viability.

It was noted, as exceptions to the general rule, that there were examples of successful introduction of new technology to developing countries. South Korea has a successful integrated steel plant based on imported ore and coal and Libya has planned a direct reduction plant based on Brazilian ore brought in carriers which will return with oil.

Paper 8

THE IMPORTANCE OF NATIONAL TRAINING FACILITIES

Case-histories for UNDP-Assisted Projects

by John H. Harris

Mineral processing, commonly regarded as an art, does in fact call upon a variety of expertise in basic chemistry and physics, in mineralogy to advanced levels, physical chemistry, thermodynamics, mathematics and, in practice, degrees of skill in economics and in chemical, mechanical, electrical and civil engineering seldom equalled, let alone surpassed in any other endeavour. Add to this the impressive array of mineral substances in use by mankind, the fact that no two mineral deposits exhibit identical characteristics, the difficulties experienced in constructing and operating mining and processing plant in the often remote and harsh environments in which minerals occur and it can be readily understood that administrative and commercial decisions in mineral engineering concepts require a grasp of scientific and economic disciplines over an unusually broad spectrum.

This has been true since early times. The legacy of endeavour which we have inherited, the objects of utility and beauty in gold, bronze, iron, ceramics and stone, the creations of architecture, the implements for husbandry and hunting, the weapons of attack and defence, all pay tribute to processes of skill and ingenuity which existed long before wholesale industrialisation began, and of which many are in use, more or less unchanged, to this day.

However, parallel with the advance of industrialisation the world has witnessed a remarkable expansion of mineral industry and of global trade in minerals. Nations in whose economies the mineral industry plays an important part have been compelled to found schools of mining and metallurgy in order to ensure a continuing supply of mineral scientists and engineers and to set up bureaux to monitor and regulate the supply of mineral products.

In addition, the need for research and development has initiated the growth of institutions, designed for experimentation in mineral processing, in both the private and the public sectors. Mining and metallurgical companies as well as the producers of mining chemicals and machinery and consultants in these fields have all contributed research and development endeavour, putting to use the basic work originating in the universities and mining schools. Not least in importance are the government-run mineral-processing research laboratories. Not only can they provide expertise, advice and guidance to the mineral industry of a country, improving existing processes and investigating new ones but they are also available as impartial advisers to government in the development of mineral policy and legislation.

To sum up, there is a recognisable need for training in mineral engineering as an aid to national development. In the past, progress in the utilisation of natural resources inevitably involved initially such basic industries as the quarrying of stone, sand and gravel, the burning of lime and the winning of salt.

At the same time legend or history usually stimulated the search for precious metals and gemstones. In the course of such random exploration occurrences of useful base-metal ores may have been discovered and perhaps coal and oil. This stage has long been passed. The inauguration of formal geological survey and the initiation of systematic mineral exploration have taken their place as pillars of national progress. As a result there have been new discoveries in sometimes unexpected localities and, as a concomitant, the need for feasibility studies which involve expertise in mineral processing and the need for trained personnel to undertake them.

The use of foreign expertise and overseas facilities may be required at first but the advantages of local establishments for testing, control and research, staffed by local graduates soon become apparent. Local knowledge and proximity of deposits permit selection of a wider spectrum of samples and avoidance of oxidation and other interfering alterations. Budgetary considerations may, however, limit the extent to which such local facilities can be equipped and operated.

In these circumstances technical co-operation has been sought by developing countries and has been forthcoming from a variety of sources, not least of which has been the agencies of the United Nations. A few examples follow.

Institution Building

In the sphere of student education, undergraduate training and postgraduate training UNESCO has executed UNDP-assisted projects in the following countries:

Egypt:	The Institute of Petroleum and Mining Engineering.
Iraq:	Field training in geology and mining.
Spain:	Technical school for mining engineering.
Turkey:	Fellowship in geology.
Tanzania:	Geological training and research.
Zaire:	The National Mining Institute, Bukavu.
Zambia:	The School of Mines.

A proposal has also been made for UNESCO/UN assistance to the Tarkwa School of Mines in Ghana. UNIDO has executed a project, which included mineral processing research, assisting the National Institute for Scientific and Industrial Research in Malaysia. All of these have components which impinge to a greater or lesser extent on various aspects of mineral engineering and on support for existing institutions. Those in Egypt and Ghana are of considerable interest since they are aimed at providing a steady flow of new technicians and engineers.

In addition there are two interesting cases of the formation, with UNDP assistance, of institutes for mining and metallurgical research, one in Bolivia and one in Chile, in which the training component has been significant.

Bolivia

The Mining and Metallurgical Research Institute (Instituto de Investigaciones Minero-Metalurgicas) was founded to aid the mineral industry and with the following objectives:

(a) Investigations in the laboratory and in the field of current methods of mineral beneficiation for the production of marketable concentrates and the possibility of improving these methods.

(b) Investigation of possible new methods, including collaboration with overseas institutions.

(c) Investigation of pyrometallurgical and hydrometallurgical methods for the extraction of metals from Bolivian ores in order to increase recoveries and establish new metallurgical industries.

The project was executed by the United Nations over a period of four years (1966–1970). Seven international experts and two consultants contributed a total of 141 man-months of varied expertise in mineral processing, metallurgy, analytical chemistry, physical chemistry and refractories technology. Subsequently supplementary assistance was provided comprising six man-months of expertise in analytical chemistry and six in chemical engineering. Fourteen fellowships (138 man-months) were awarded covering training in mineral processing methods and design, ferrous and non-ferrous metallurgy, mineralogy and mathematics. The national staff was built up to a total of 60 engineers of whom six had the degree of Ph.D. They were drawn from Bolivian mining schools. Many had had their training re-inforced by studies abroad and by operational experience outside as well as within Bolivia. Research programmes were undertaken which received wide recognition through presentation of papers at international congresses and symposia. Development work and industrial investigations were conducted on bench and pilot scale, including the setting-up of pilot-scale flow-lines within operating plants. In contract work for mining companies representatives of the companies were encouraged to take part in all phases of the investigations. Training of engineers, technicians and operators was thus wide-spread. In addition leading members of the staff of the Institute lectured at the local University and all of the staff were encouraged to lecture on their work at the Institute's weekly internal seminars. Furthermore, formal courses were run for training analytical chemists for the mining industry, for training computer operators and in the English language. The Institute also provided numerous opportunities to students for vacation work both in the laboratories and in the field. At the present time (1975) the Institute continues in a flourishing condition and is seeking further inputs of expertise from the United Nations.

At the same time one of the national staff has now progressed to the status of international expert and will be employed by the United Nations in the new project in Southeast Asia referred to below.

Chile

The Mineral and Metallurgical Research Centre (Centro de Investigacion Minera y Metalurgica) was set up in 1971. UNDP assistance was first programmed to September 1974 and has since continued into a second phase which is due to last until the end of 1977. The Government of Belgium participated to the extent of a contribution of $900,000 for buildings and $47,000 for equipment. The project is being executed by UNIDO in association with the United Nations.

The purpose of the project was to design, build and equip a comprehensive research and development centre to serve the mining industry of Chile in general and, in particular its advanced and extensive copper industry. The Centre has been developed to include departments for research in mining, mineral processing, mineralogy, analytical chemistry, pyro and hydrometallurgy, electrometallurgy and pipe-line transport of concentrates. It has, as one of its main objectives, close collaboration with universities in Chile and was temporarily accommodated in university departments. Now (1976) it has moved into its own buildings which are nearing completion.

Training has been an important feature of this project, which provided for the following fellowships:

	Phase I	*Phase II*
	Man-Months	
Materials characterisation	18	27
Mining	30	48
Process metallurgy	48	24
Process engineering	24	30
	120	129

The work of the international team of experts and consultants (144 man-months in Phase I and 340 man-months in Phase II) contributes to training on the ground.

The Centre aims to provide for graduates a thorough grounding in industrial research and so ensure a continuing supply of recruits to the mining industry. At the same time it provides a focus from which new techniques can be disseminated and in which experienced engineers from industry can meet to discuss problems and to develop solutions in collaboration with trained researchers and with modern equipment. Thirdly, and not least important, it provides teams of researchers and pilot-scale equipment for field-work in prospects, mines, mills and smelters in actual operations.

Important results have already been achieved in the fields of mining, beneficiation, smelting and refining of copper ores in which the Centre is bound to specialise and (like its counterpart in Bolivia, which specialises in tin) aim at international recognition.

Southeast Asia

At the request of Indonesia, Malaysia and Thailand, producers of the major part of the world's tin, the UNDP is assisting in the establishment of the Southeast Asia Tin Research and Development Centre, a regional project for which Malaysia will be the host country. The region is, of course, famous for its alluvial tinfields while relatively little development of underground mining has taken place. The objectives of the Centre will be to improve efficiency and productivity in the existing regional tin-mining and smelting industries and to systematise the exploration for and development of underground primary sources using, for these purposes, expertise developed in other regions. Fellowships will be provided for study overseas; international experts and consultants will lead the researchers for the first few years and provide training locally.

The Centre will work in collaboration with the International Tin Council and form a regional counterpart, at the mining end of the Council's Tin Research Institute which deals with development of the uses of the metal. It is also intended to become a regional centre for the collection and distribution of technical and economic information and for seminars and symposia concerned with the tin industry.

Reinforcement of Existing Institutions

In this sphere UNDP assistance has been varied and wide-spread in recent years. This type of project usually deals with a specific development objective in an existing operation and, while fellowships are often included within the terms of the assistance, they too are directed at the acquisition of limited precise skills rather than general training. Most of the training is accomplished on the ground during the pursuit of the objective, with the assistance of UNDP-provided equipment and international expertise. The national institution, at the end of the project, is then better armed as a provider of national facilities.

Burma A great deal of assistance was given to improving operations at the Bawdwin Mine. More recently improvements were put into the laboratories of the Directorate of Geological Survey and Exploration and a demonstration gravel-pump mine and dressing plant were installed for the Myanma Tin and Tungsten Corporation. The Myanma Mineral Development Corporation has now requested help with its planning and processing division and with a feasibility study for a copper smelter.

Burundi Following on the discovery of deposits of nickeliferous laterite by a UNDP-assisted mineral survey the government has requested a feasibility study of the possibilities for production. Provision will be made for training of nationals in all phases of the study.

Ghana The State Gold Mining Corporation has received assistance in training in the planning and prospecting of alluvial gold mines. The establishment of comprehensive training facilities in a large underground gold mine is also in hand.

Indonesia The project here was concerned with two aspects of offshore mining for alluvial tin. One was the improvement of methods of prospection and evaluation, the other was the improvement of processing methods for the recovery of offshore tin and its associated minerals. Both aspects involved innovations in plant design, construction and operation and the prior training abroad of national engineers. In operation, however, a strong element of training was dominant. The project was operated in conjunction with the State Tin Enterprise which, as a result, achieved the possibility of providing vacation training for mining students and research projects for their theses.

Jordan This project, recently commenced, will provide research facilities on bench and pilot scale for the Jordan Phosphate Mining Co. Training of engineers overseas has been accomplished. The new laboratories will provide facilities for research into new methodology for beneficiation of low-grade and complex phosphate ores and a training ground for new entrants to the industry.

Pakistan Here again the government has requested help with a feasibility study for a copper mine, beneficiation plant and smelter. Such a project will also be designed to ensure the evolution of a permanent training facility.

Southeast Asia A regional training course in uranium prospection and the evaluation of ore deposits was run by the International Atomic Energy Agency.

Turkey The Institute of Mineral Research and Exploration (Maden Tetkik ve Arama Enstitüsü) (MTA) has a great responsibility for assisting in the training of new entrants to the field of mineral engineering. Part of the recent UNDP technical cooperation was directed to improvement of the facilities of the mineral processing research laboratories and to solution of some of the beneficiation problems of complex domestic ores. The potential for training has thereby been enhanced. A programme of further cooperation is at present under consideration by the MTA.

Protection of the Environment

Mining, by definition is "to get by digging". As a consequence of mining there must therefore be disturbance of the environment. This brings out the necessity for training mineral engineers, the public and their governments into an awareness of the problems involved and the means available for their alleviation. The mineral industry, with its mines, beneficiation plants and smelters is not the only disturber. Agriculture, forestry, fishery and manufacture can all contribute their pollutants, their effect on run-off and drainage and their erosion of biological systems. Mining may, however, be at a disadvantage in that, although it occupies only a small fraction of the land surface of any country, it may compete for land urgently needed for other occupations. Wise decisions on priorities are required and repair of the disturbance is often indispensable.

To these matters the UN gives practical consideration through the United Nations Environment Programme (UNEP). In the field of mineral engineering three current projects may be mentioned.

Jamaica The Government of Jamaica, in collaboration with partners, plans to institute a national bauxite mining and alumina production industry. The mining of bauxite inevitably interferes with agriculture and involves relocation of populations. Bayer process alumina plants introduce pollutants which are difficult to contain and usually sterilise hundreds of acres of land under tailings. Transport of finished product to the coast and its export through newly constructed port facilities inevitably have an effect on the vulnerable ecology of a coastline. The government has therefore set up an environmental impact study in which mining and metallurgical engineers will work together with terrestrial and marine biologists, oceanographers and coastal protection specialists. UNEP will assist this operation, designed to define in advance the methods of alleviation of the problems to be encountered.

Papua New Guinea The Bougainville copper mine, one of the largest and richest in the world has had a tremendous impact on the biological background and especially on the human environment in that small and relatively backward island. The mining company maintains an environment protection division and the government, with UNEP assistance, is forming a permanent environmental audit department. The objective is to derive experience of control of environmental disturbance for use elsewhere, in view of the likelihood of further large-scale mining developments in the country.

Southeast Asia These countries offer typical examples of the way in which surface mining competes with agriculture for scarce land. Constraints based on imperfect technology and adverse economics have the effect of creating situations where it is either impossible to reclaim the mined-out land or inadvisable to reclaim it because un-recovered values remain and retreatment at a later date is foreseeable. UNEP is conducting an investigation into the possibilities of increasing the rate of reclamation, including simultaneous reclamation concurrent with improved mining and recovery techniques.

Conclusion

It might be considered appropriate to attempt a definition of the meaning of training. In this there is, however, as much difficulty in definition as in implementation. Transfer of technology may not be sufficient in itself and in many an instance the teacher becomes the taught and the student the master. Language may be a barrier to the transfer of knowledge. However impressive the possibilities of theoretical and practical training in established institutions overseas the advantages can well be seen of training in the national environment. In this, the developing countries and the UNDP work with a synergistic purpose—the simultaneous action of separate agencies which, together, have greater total effect than the sum of their individual effects.

Discussion Summary

In discussion education and engineering training were stated to be fundamental for the development of an economy by industrialisation, but lack of entrants into mineral engineering courses, lack of teachers, low salaries and status for civil service employees and consequent drain to industry were difficulties. There has been a drift away from the attractions of engineering to those of social philosophy and computer science but it was reported that in France the trend was being reversed by higher salaries and status, at any rate for mineral engineers. It was asked whether it was desirable to set up specialised training centres in developing countries or to send students away to advanced countries. It was agreed that local training centres of suitable types were desirable. Countries could co-operate on a regional basis to make most use of existing or planned facilities such as the new drilling school in Ghana and existing UN regional projects in Kuwait, Yugoslavia and Guyana. There was comment that the UN system of fellowships appeared to vary in terms and numbers offered. It was suggested that, in the countries concerned the United Nations Resident Representative should be consulted about the funds available and that a visit by a Technical Adviser from UN Headquarters could be requested to assist in formulation of fellowship and other training programmes.

Paper 9

ENVIRONMENTAL FACTORS AND THE REQUIREMENTS FOR CONTROL OF EFFLUENTS

by J. Lajzerowicz

In this paper the importance of minerals in the economic growth of Canada will be outlined. The variety and importance of different kinds and sources of environmental damage and its associated problems will be presented. This will be followed by a discussion of the main issues related to the economics of environmental control as well as the definitions and estimates of pollution control costs. Also, highlighted will be the main elements in the philosophical framework for the environmental protection policy in Canada. The paper will end with a brief indication of the way in which these policies are being applied to mineral development.

Canada's balance of trade is an important factor in the nation's economic well being. In this, the mineral industry has a vital role, for more than 60% of Canada's mineral output is being exported. This represents about 25% of total exports. In the world context, Canada (1973) ranks first as a producer of asbestos, nickel, silver, zinc and titanium; second in potash, gypsum, uranium, sulphur and molybdenum; and third in platinum, gold and copper. Its share of the world's production of some selected commodities is shown in Table 1.

Table 1.

Commodity	Year 1950	1970	2000
Copper	10%	10%	11%
Nickel	80	42	30
Iron Ore	1	6	5
Lead	8	10	8
Zinc	9	23	20
Asbestos	65	43	45
Potash	0	18	22

Source: Department of Energy, Mines and Resources, Ottawa.

Growth in production has been stimulated by foreign, mainly American-demand. This has resulted in Canada becoming the world's third largest diversified mineral producer. Moreover, it will continue to be a major pro, ducer for years to come as suggested in table 2 showing estimated reserves for some commodities.

Table 2.

Commodity	1974 Output in short tons	1974 Domestic Consumption as percentage of 1974 Output	Canadian reserves (measured and indicated) of Metal Contained in Mineable Ore as of January 1, 1975* Producing mines and deposits under development	Other deposits for which production is foreseen
Aluminium	1,125,329	35.2	—	—
Asbestos	1,811,938	2.4	—	—
Copper	905,417	30.2	18,790,900 s.t.	12,929,900 s.t.
Gold	1,718,000	—	11,911,500 oz.	604,500 oz.
Iron Ore	51,571,084	24.9	4,704,000,000 s.t.	45,300,000,000 s.t.
Lead	324,375	24.2	10,261,700 s.t.	2,562,200 s.t.
Molybdenum (Mo content)	15,368	12.0	757,600,000 lb.	878,215,200 lb.
Nickel	296,600	4.3	8,011,700 s.t.	669,600 s.t.
Potash (K_2O equivalent)	6,366,971	3.5	2,385,600,000 s.t.	
Silver	43,765,000	24.4	919,267,300 oz.	102,666,200 oz.
Zinc	1,242,314	10.4	31,166,200 s.t.	7,315,400 s.t.

* D.A. Cranstone and S.A. Hamilton, "1975 Canadian Reserves of Seven Metals," *Canadian Mining Journal*, February 1976.

While presenting Table 2 it must be pointed out that the key to future rates of mineral resource development is not found in tables comparing reserves and consumption but in preferences and ethics[1].

The pressures of a new world economic order as well as Canada's national aspirations have led to a review of mineral policy. The discussion now taking place on how minerals can best contribute to social and economic progress is centred around the following four policy options:[2]

To continue, as in the past decades to encourage maximum mineral production.

To encourage economic diversification and growth through increased mineral processing and mineral-based manufacturing in Canada.

To obtain the highest possible net financial returns to Canadians from minerals.

To conserve mineral resources for long-term domestic requirements.

It is not the purpose to elaborate on these options here. It is not likely that "business as usual" plus technological advance will become *the* strategy: rather, the policy will be one which favours better regional distribution of mineral activities within Canada, encourages secondary industry, seeks increased Canadian ownership in the resources sector, and recognises the need for effective competition in international markets. The latter is necessary simply because the domestic economy is not large enough to absorb the production required to achieve economies of scale.

Environmental Impact of Mineral Development: Status and Problems

Jeremy Bentham in one form or another but again and again said that "nature has placed mankind under the governance of two sovereign masters, pain and pleasure. It is for them alone to point out what we ought to do, as well as what we shall do."[3] So it is with the aftermath of mineral production. On the one hand, minerals bring prosperity; on the other, they bring problems.

Industrial civilisation would be impossible without minerals, yet the activities connected with mineral development degrade the environment. Our economic system, with its expanding production, increased use of resources and discard of wastes has come to be viewed by many as responsible for both the depletion and the pollution of the environment. Concern for the damage done by pollution depends in part of what effects people consider to be obnoxious. Individuals, as well as generations, differ in their perceptions. To a large extent, pollution is what people consider it to be, but by and large these effects are a by-product of our way of life and "pollution control" is ultimately tied to the political process. We should however realise that pollution is also in a narrower sense a measurable cost which does not depend on perception.

Impacts on the environment resulting from mineral activities differ depending on the stage of the mineral system. They affect man at all stages. With respect to the natural environment the major impact during the exploration and mining stages, is on land and on water. At the processing stage however, it is air and water pollution which become the major problems. Details concerning the pollutants and effects in different areas of mineral activities are shown in the matrices on pages 188–189.

It is interesting to note that the Canadian mineral industry in 1970 contributed 42.8% of the particulate matter and 63.8% of the sulphur oxide emitted into the air.[4] While major improvements are being made in the technology of production to reduce effluent per ton of output, in transport facilities to spill less per ton-mile etc. we have not yet been able to come up with a significant improvement in the environment. This is not just because of industry's reluctance to spend more money but because the decrease of effluents per unit of output or of consumption, must more than overcome the increase in production or consumption itself. It should also be noted that in many instances the damaging effects of pollution on the natural environment can continue in some cases for hundreds of years even though the release of a pollutant is being stopped. As well the effect of pollution on man is often not immediately evident. For example, on the average, silicosis symptoms have taken 30 years to appear in foundry workers and 20 years in gold miners. Lung cancer among uranium miners took between 3 and 8 years to mature. In addition, we still lack precise knowledge on monitoring methods, "tolerance levels" and means of control.

To illustrate the evident damage to the natural environment resulting from mineral activities, let us have a look at the operations of the biggest producer of nickel in the world and of copper in Canada—International Nickel Company, operating in Sudbury, a city of over 100,000 inhabitants. By 1970 the smelter stacks were emitting almost 3 million tons of SO_2 per year, and damage to the soil and aquatic life could be detected over an area in the excess of 2,000 sq. miles with the most damage occurring in the central core, covering around 600 sq. miles. In 1972, a 1250 ft. super stack was constructed in an attempt to dilute fallout in the Sudbury area. However, this appears to have spread pollution over a wider area and it is estimated that acidic particulate matter might now be carried distances of up to 250 miles in sufficient concentrations to cause concern. In addition to SO_2, metals emitted in the Sudbury region are also more widespread and at higher concentrations in the soil than in all other known instances.[5] It is still not known how much of the damage to both soil and vegetation can be attributable to metal contamination and how much to SO_2 emissions. It is known however, that out of 30 lakes in the area once known for their fishing, six now have no fish at all, 18 have no lake trout and the rest have all shown marked signs of decline.[6] It is well documented in the literature that excessive SO_2 is definitely hazardous to human health as well as animal and plant life and experience shows that revegetation often does not occur naturally within 70 years of the closure of a smelter.

Generally speaking much is known about the effects that pollutants have on man, animals and plant life but even more is still to be learned. Between 1954–1969, two companies mining gold in the Northwest Territories, Canada emitted into the air and water over 8,000 tons of arsenic. The significance of this discharge cannot yet be fully evaluated. There is however growing suspicion that it may cause chronic bronchitis, neurological abnormalities and skin lesions. It has also been suggested that there is an association between exposure to arsenic and the development of cancer[7].

By giving the above examples, it has been intended to illustrate the problem. It is not the purpose to imply that mining companies are to be blamed for it. In fact, as Henry Ford II has said, "there are severe limits to what business can do, entirely by itself, to solve environmental problems. The reason is that in the absence of appropriate government regulations, there is only a limited market for the main solutions to environmental problems. Few consumers, for example, will pay much for lower vehicle emissions unless their choices are limited by government regulation. Business lives and dies in the market, and no company can survive if it voluntarily assumes pollution control costs far out of line with those of its competitors."[8]

It has required some time for us to realize the extent of the problem which has only recently taken serious proportions. According to most observers there are three principle explanations for the emergence of environmental problems—the increase of population, the spread of affluence, and

Matrix 1 ENVIRONMENTAL IMPACT OF PRIMARY STAGES OF MINERAL PRODUCTION

AREA OF ENVIRONMENTAL IMPACT		PRIMARY STAGES OF MINERAL PRODUCTION			
		EXPLORATION	MINING	BENEFICIATION & STORAGE	TRANSPORTATION
HUMAN RECEPTOR	Pollutants		UNDERGROUND: Wet, Dust, Noise, Radiation, Heat, Gases OPEN PIT: Dust, Noise, Exhaust Fumes.	Dust, Noise. NON-FERROUS: Noxious Fumes, Handling Problem with some Toxic Reagents. ASBESTOS: Asbestos Dust.	Noise, Dust, Exhaust Fumes from Heavy Vehicles.
	Effects		Contributes to Labour Turnover. Possible Respiratory Disease, Particularly for Asbestos, Fluorspar, Gold. Stress & Other Physical Ailments. Cancer.	Contributes to Labour Turnover. Possible Respiratory & Carcinogenic Diseases, Particularly for Asbestos & Other Dry Milled Minerals.	Possible Respiratory Effects During Handling.
	Corrective Measures		Environmental & Safety Training of Staff for Good In-House Practices, Respirators & Other Protective Clothing, Good Ventilation, Dust Collectors & Inhibitors. OPEN PIT: Dust Repressant Sprays.	Environmental & Safety Training of Staff for Good In-House Practices, Proper Protective Equipment, Greater Automation Dust Collectors, Vacuum Equipment, Good Ventilation.	Unit Trains & Distribution Centres for Aggregates Truck Routes. Zoning Plans. Special Handling Equipment, Dust Collectors, Covered Cars, Spraying.
LAND ECOSYSTEM	Pollutants Disturbance	Trenching, Drilling, Access Routes, Line Cutting, Abandoned Equipment.	OPEN PITS & QUARRIES: Tailings Disposal, Clearing of Overburden & Vegetation. Roads & Access Routes. Waste Piles. Subsidence. Impact of Poorly Designed Townsites.	TAILINGS DAMS & PONDS: Dust, Contamination from Seepage & Spills. Unsightly Stockpiles (eg. Sulphur). Waste Dumps (eg. Salt Tailings) from Potash Production. Alumina Production - Red Mud Waste.	Wide Roads for Heavy Vehicles & Associated Borrow Pits. Dust. Unnecessary Clearing. Spills from Derailments & Road Accidents. Arctic - Problem of Road, Port, Airstrip Construction.
	Effects	Erosion, Scarring (Minor), Vegetation Damage, Permafrost Damage in Arctic, Disruption of Natural Drainage.	Limited Sequential Land Use Possibilities, Health Hazard from Contaminated Water. Aesthetics. Permafrost Damage in Arctic.	Waste Lands Created by Tailings Areas. Land Contamination by Leaching & Run Off from Tailings and Waste Dumps. Permafrost Damage in Arctic.	Opens Wilderness Areas to Possible Degradation. Heavy Traffic Destructive to Highways. Permafrost damage in Arctic.
	Corrective Measures	Reclamation, Plug Drill-Holes Where Necessary. Proper Design of Access Roads. Generally Good Housekeeping Practices.	Reclamation, Proper Design of Tailings Areas (Conservation of Top Soil, Revegetation, Backfilling, etc.). Possible Sequential Land Use. Use of Waste Material (eg. Construction Aggregates). Reduce Subsidence by Fill Methods & Proper Planning of Pillars.	Stabilization of Tailings Area (eg. Revegetation, Landscaping). Proper Dam Construction & Operation. Impermeable Dams, Downstream Ponds, Drainage Water Recycle. Zoning of Operation with Respect to Urban Areas (eg. Aggregates, Cement Plants, etc.) Dust Control.	Planning Road Design to Minimize Ecological Disturbance. Maximum Utilization of Existing Roads. Integrated Regional Road Networks.

WATER ECOSYSTEM	Pollutants	Suspended Solids from Erosion. Salt Exploration: Brine from Drilling into Aquifers.	Suspended Solids from Mine Water, Seepage from Tailings Areas, Heavy Metals, pH from Metal Mines, Alteration of Water Table.	Suspended Solids, Heavy Metals, pH, Toxicity from Direct Discharge & Overflow, or Seepage from Tailings Systems. POTASH & SALT MINES: Salt Contamination of Water Bodies.	WATER TRANSPORT: Suspended Solids**. Colouration (Iron Ore) at Shipping Terminals. Spills from Derailments & Road Accidents. Possible Problem from Slurry Pipelines.
	Effects	Contamination of Streams & Ground Water.	Harmful to Water for Human Consumption and to Aquatic Life.	Harmful to Aquatic Life. Food Chain Effects. Possible Hazard to Human Health.	Possible Harm to Aquatic Life.
	Corrective Measures	Reclamation. Plug Drill Holes Where Necessary.	Treatment for Mine Water. Good Design of Tailings Areas.	Mill Water Recycle or Treatment for Discharge. Well Constructed Tailings Dams. Improved Operating Procedures.	Care in Handling. Special Loading Equipment. Better Railcar & Track Maintenance. Treatment or Recycling of Pipeline Water.
			Recycling Mine Water in Milling in Integrated Operations*		
AIR ENVIRONMENT	Pollutants		Wind Blown Dust. Gases from Fuel Burning Equipment. Dust from Drilling, Blasting.	Dust, Airborne Particulates. Gases, Odours, Evaporation from Tailings Ponds. SO_2 from Concentrate Drying, Heavy Metals. Thermal Power Generation: Hydrocarbons, SO_2, NO_x.	Airborne Particulates from Loading Material Being Transported & from Road Surface.
	Effects		Minor	Possible Respiratory Effects. Rain Washing of Particulates Affects Vegetation and Soil. Increased Costs from Corrosion, Dirt, etc. If Close to Urban Areas, Health Effects from Asbestos Fibres.	Minor
	Corrective Measures		Dust Collectors & Sprays Where Necessary.	Collection Systems for Dust & Particulates. Stabilization of Tailings Dumps.	Covered Shipments or Sprays Where Necessary. Stabilization of Road Surface. Dust Collection on Handling Equipment.

Matrix 2 ENVIRONMENTAL IMPACT OF SECONDARY STAGES OF MINERAL PRODUCTION

AREA OF ENVIRONMENTAL IMPACT		MINERAL PRODUCTION ACTIVITY			
		IRON & STEEL MAKING	NON-FERROUS METALS, SMELTING & REFINING		NON-METALLIC MINERALS, FABRICATION
			Copper, Nickel, Lead, Zinc & Others	Aluminum	
HUMAN RECEPTOR	Pollutants	Dust, SO_2, CO, Hydrocarbon Derivatives, Heat, Noise, Acid Mist.	SMELTING: SO_2, Gases, Heavy Metals, Particulates, Noise, Lead Fumes. REFINING: Acid Mist, Fumes, NH_3, H_2, Nickel Carbonyl.	SMELTING: Particulates, Gases (Fluorides), Noise, Heat, High Electro-Magnetic Fields.	Dust, Noise. ASBESTOS: Fibre & Dust Problem.
	Effects	Possible Respiratory Effects & Lethal Accidents, Cancer.	Respiratory Diseases, Cancer.	Possible Respiratory & Other Diseases. Effect of Magnetic Fields Unknown.	Respiratory Diseases & Cancer.
	Corrective Measures	Dust Collectors, Scrubbers. Adequate Ventilation Systems.	Gas Collection & Treatment. Protective Equipment. Adequate Ventilation Systems.	Dust & Gas Collection. Adequate Ventilation Systems.	Adequate Ventilation Systems. Face Masks. Vacuum Equipment.
LAND ECOSYSTEM	Pollutants Disturbance	Accumulated Slag & Metallic Oxide, Dust, Sulphide Leaching of Slag.	SO_2 & Particulate Rain from Stack Plumes, Slag Heaps.	Miscellaneous Waste Materials.	Unsightly Product Storage Yards, Dust.
	Effects	Contamination by Leaching. Aesthetic Effects of Waste Dumps & Areas.	Aesthetics. Burning of Trees & Vegetation, Soil Contamination by Heavy Metals. Corrosion of Materials.	Aesthetics.	Aesthetics, Limitations on Land Use.
	Corrective Measures	Use Slag for Land Fill. Iron Dust can be Pelletized & Recycled, Dust Control, Reclamation.	Use Slag for Land Fill and Road Bed Aggregates. Stack Gas Treatment. Process Modification, Reclamation.		Zoning for Plants, Dust Control, Reclamation.

WATER ECOSYSTEM	Pollutants	Pickling Acids (HC_1, H_2SO_4, H, HNO_3), Sludges, Ether Solubles. Suspended Solids, Bod, Phenols Cyanides, Heavy Metals, Phosphates. Heat Pollution, NH_3, Leaching from Slag.	SMELTING: Suspended Solids from Slag, Leaching, Acid Rain, Mercury. REFINING: Suspended Solids, pH, Toxicity, Leaching & Electrolytic Waste Solutions, Heavy Metals, Leaching Solutions.	Suspended Solids, Fluorides, Ether Solubles.	Dust, Leaching.
	Effects	Harmful to Aquatic Life, Possible Pollution of Potable Water Supply.	Harmful to Aquatic Life, Acid Build-Up in Lakes & Other Toxic Substances (e.g., Heavy Metals). Can Affect Potability of Water.	Harmful to Aquatic Life.	
	Corrective Measures	Recycle HC_1. Eliminate H_2SO_4. Cleaning of Steel. Suspended Solid Removal Systems, Pollution Control Equipment, Chemical Treatment, Process Changes.	Process Water Treatment & Recycling. Recover SO_2 & Particulates from Stack Gases. Process Change.	Eliminate Wet Scrubbers & Dry Scrub Fluorides with Alumina Packed Bed Filters.	Transport of Ponded Solids to Suitable Disposal Areas. Develop Uses for Waste Materials.
AIR ENVIRONMENT	Pollutants	COKE OVENS: SO_2, Particulates, Hydrocarbons - Corrosive Gases & Dust, Airborne Dust - Noxious Gases & Dust. BLAST FURNACES: H_2S from Slagging, Particulates, Steel Making: Particulates, Fluorides, Cd, CO, Acid Fumes, Hydrocarbons, SO_2.	SMELTING: SO_2, Particulates, Heavy Metals (As, Cd, Hg, Pb, Zn). Lead: Lead Fumes, Mercury. REFINING: Particulates, Fumes. Zinc: Acid Fumes, $ZnSO_4$ Fumes. Nickel Hydrometallurgy: NH_3.	SMELTING: Particulates, Fluorides, Hydrocarbons, SO_2. REFINING: Particulates, C_1, SO_2 (Drying Fuel)	Particulates, Asbestos Dust.
	Effects.	Corrosion of Metal Surfaces & Building Facades, Blackening, Paint Deterioration.	Respiratory Diseases, Transport of SO_2 Pollution, Vegetation Damage.	Fluorides Toxic to Some Vegetation, Direct Effect on Humans, Food Chain Accumulation of Fluorides.	Minor Health Hazards.
	Corrective Measures	Gas Scrubbing at Source, New Types of Coke Ovens, Process Change, Material Substitution.	Recover SO_2 from Stack Gases, Bag Houses, Electrostatic Precipitators, Scrubbers, Zoning of Plant with Respect to Urban Development, Process Modification.	Dry Scrub Fluorides.	Dust Collectors.

the greater impact of technology. None of these necessarily excludes the others: they could be cumulatively and jointly responsible.[9]

Maurice Strong, former Head of the United Nations Environment Programme, presented[10] very interesting statistics characterising the scale of the revolutions through which we are now living:

It took one million years to produce the first billion people on earth. It will now take approximately 15 years to produce the next billion.

From the beginning of civilised life on earth to the end of World War II, all of the world's industry totalled less than the new industrialised capacity which has been produced within the past three years.

In one decade, 1959 to 1968, the United States alone used more resources than all the world's people in all of previous history.

In primitive society, man consumed the energy equivalent of approximately 220 pounds of coal per person per year; the average per capita use of energy in industrialised societies like the U.S. and Canada is about 11 metric tons of coal per year, and world energy conversion is growing at almost three times the rate of population growth.

The Economics of Environmental Control

It is quite clear that Canada, as well as other industrial nations, is proceeding beyond the traditional industrial age. The fundamental assumption now is that air and water are also natural resources and should not be squandered as though they were unlimited in quantity. They have to be appreciated as scarce resources requiring economic use like any other resource.

"In some cases, depletion and environmental change are necessary and acceptable; in others, one or the other or both must be regarded as waste and degradation; and in most cases, either improved recovery of minerals requires a greater impact on the environment, or protection of the environment dictates lower levels or rates of mineral recovery. The analytic questions are when and where and for whom each case applies, and these questions can only be answered in terms of what economists call tradeoffs. To evaluate these, one must compare what has been gained with what has been lost because of the production or use of minerals. Such difficult comparisons must include physical units as well as economic, and in many cases social units as well".[11]

In other words material policy is intimately intertwined with social and economic policy. Unqualified further material growth, which is still seen by many as an admirable end in itself, will likely lower environmental quality.

In the technological tradeoffs involving the environment there are very little Canadian data available. However, it is reasonable to state that we now face a situation in which widespread use of new technological developments may occur before their social impact can be properly assessed and long before any empirical adjustment of the benefit-versus-cost relation can obviously be indicated. This fact comes as no surprise. It is an extension of the *problem* of mineral development once borne almost wholly by the Native Peoples of Canada (Indian and Innuit alike). While the impact on their traditional livelihood has been measurably disastrous due to activities in more than just the mineral field, today, with government intervention, it has been somewhat eased. Through consultation with governments, negotiation and, if need be, action in the courts, Native Peoples are assuring themselves that more proper pre-assessment will take place and techniques developed that will reduce environmental damage to a minimum.

There are many examples, especially from developing countries, where major industrial projects are already leading and promise to continue to lead to a wide range of environmental problems. A case in point is the Cuajone copper mining and smelting project in Peru. The project, which is located in a sparsely inhabited and largely semi-arid area immediately adjacent to the Pacific coast in southern Peru, will produce 186,000 tons of blister copper annually. It is estimated that 30 million tons of tailings will be discharged into the sea and 60,000 tons of sulphur oxides will be emitted into the air annually. Consequently, population centres and agricultural areas are already alarmed at the prospect of being exposed to such pollution and who can blame them? Without adequate environmental safeguards the project could have a far reaching impact on Peruvian agriculture and the world's largest anchovy fishing industries, to say nothing of the health of the residents in nearby urban centres.[12]

Cost-benefit analysis can provide us with an objective assessment, but while we know how to analyse costs, it is extremely difficult to bring into the calculus every benefit and every cost associated with a particular project.

The benefit side has been very neglected and public policy cannot possibly be set without explicit or implicit benefit estimates. The claim of some that only economic solutions can be applied to pollution control is questionable. In principle, it is possible to determine the amount of tolerable pollution but in practice it is often difficult to measure. Consequently, if it is immediate action that is required, then a value and not an economic judgement will have to be made. After such a decision has been taken, the economic side can be introduced to ensure that the offending pollutant is efficiently controlled. Not surprisingly the outcome of such subsequent cost-benefit analysis will most likely vary according to the economist in charge due to the judgmental factors as to what is to be included and how it is to be evaluated.

The subject of how to approach environmental costs is being widely discussed. There are serious difficulties in preparing and using environmental cost estimates, but despite these, estimates must be attempted if society and the policy maker are to have the needed information to adequately understand the trade-offs involved.

The following is a diagramatic summary of the concept prepared by

Environment Canada[13] in 1974, based on the research of J.H. Dales[14] and the EPA[15].

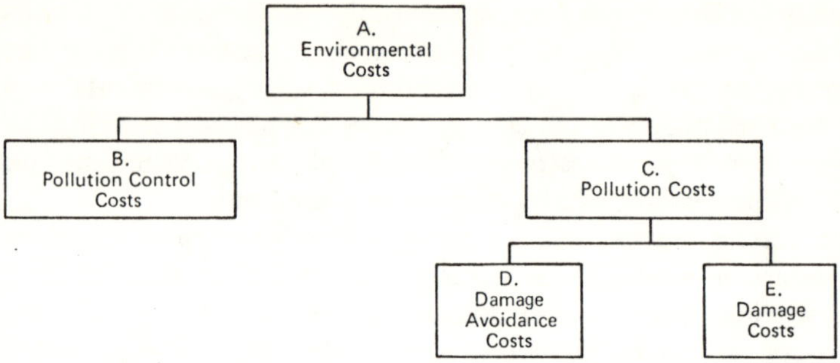

A. — The total costs incurred by society in the managing of its wastes (includes costs estimates in terms of air, water, solid wastes as well as noise pollution, aesthetic degradation, etc.).

B. — Refer to amount of money spent by public bodies or private parties to *prevent* pollution *per se* and not the cost preventing merely the harm that pollution causes (eg. treating of sewage by municipalities).

C. — Refer to the money value of the damages caused by wastes after they are released in to the environment.

D. — Expenditures to prevent pollution damage (incurred either by the government or by private parties). eg. the treatment of municipal water supplies.

E. — Money equivalent of the reduction in welfare resulting from pollution damage that is not prevented (eg. the loss of swimming facilities due to the closing of local beaches on account of pollution).

There are innumerable estimates of how much it will cost to control pollution in this decade. Few agree, and it is difficult to find out which costs have been considered and what level of pollution reduction has been achieved.

A study prepared for the Organization for Economic Cooperation and Development (OECD)[16] concluded that "the costs of present programmes for additional pollution control are manageable but need managing". Relevant data from this document are summarised in Table 5.

Table 5.

Country	GNP per capita U.S. $	Population density per km²	Annual Pollution Control Costs as per cent of GNP[17] 1975	1980
U.S.A.	4,816	22	0.8	1.4
Germany	3,089	240	0.5	—
Sweden	4,055	18	0.6	—
Netherlands	2,410	319	0.4	1.2
U.K.	2,120	228	—	0.4
Italy	1,794	178	0.3	1.0
Japan	1,921	280	3.3	—

Some data covering costs of pollution abatement specific to the mineral industry is shown in Table 6. The estimates given are based on the assumption that the level of effluent treatment involved represents "Best Practicable Technology".[19]

Table 6.

Commodity	Pollution Abatement Costs $/lb. Minimum	Probable	Maximum	Approximate selling price	Probable Cost as % of selling price
Aluminium	0.020	0.025	0.032	0.30	8
Copper	0.022	0.025	0.050	0.60	4
Lead	0.012	0.014	0.017	0.14	10
Zinc	0.0123	0.0123	0.0267	0.18	8
Iron & Steel	0.00375	0.00460*	0.00545	0.15	3

Source: The Economic Impact of Pollution Control, CEQ, 1972.
* Arithmetic average.

Canadian costs are fairly comparable on a per capita basis with those of the US, one estimate[19] for air pollution control being $52 per capita in 1965. Assuming a 4% per year rate of increase and using present population of 23 million, the gross total would be about $1.8 billion in 1975.

The bulk of this expenditure is presently being borne by those in the largest commodity sector—in terms of numbers of operations—the copper and nickel producers. The Mining Association of Canada has indicated that capital costs required for environmental control purposes in new mineral developments now average about 10–12% of total capital outlay (some of this amount would have to be spent under any circumstances, simply to get rid of wastes). In some cases this cost may be less, but is likely to continue to be the trend in new installations.

This is a significant added cost and one which ultimately will be absorbed one way or another by the consumer. On the other hand these costs will have to be absorbed into the development package of any new mining venture and in the long run Canada's environment will be that much cleaner.

Environmental Regulations: Trends and Impact on the Mineral Industry

Canada is among the most active countries advocating a controlled and balanced interface between mineral development and the natural environment; although it is recognised that the subject of environmental control is still in many ways more of an art than a science. The situation is complicated in Canada by the fact that there are both federal and provincial jurisdictions involved. When put together with regional economic disparities that exist across the country, these factors accentuate the problems arising from the conflicting objectives of increased industrialisation and the preservation of a pure environment.

In the past, legal action and the regulation of emissions generally occurred after the damage to the environment was done. The trend today is to ensure from the start that environmental damage is kept within acceptable "limits". This can be affected by various economic incentives and regulations.

Under present constitutional arrangements between the federal and provincial governments, ownership of natural resources is under provincial jurisdiction. Thus, control of pollution resulting from mineral activities has a large measure of provincial responsibility. The federal government's role (through the Department of the Environment) is to administer a number of acts—at present 22—the focus of which is clearly on the renewable resources, the quality of the natural environment and the provision of information and knowledge. The Department coordinates federal environment programmes, co-operates with the provinces and has responsibility for environmental problems having interprovincial or international implications.

Environmental controls at both federal and provincial government levels have been formulated to take account of conservation practices, operations guidelines, offences and penalties, damage restoration, and the general rights, duties and powers of authorities under environmental legislation. The direction of recent legislation has been toward the over-all management of resources and their exploitation rather than the more segmented proscriptive approach of the past. New environmental management legislation now exists for all provincial and federal jurisdictions. The bulk of its formulation has occurred since 1970, along with the provincial administrative trend to establish their own departments of environment. The fundamental principle that the polluter pays is inherent in Canadian policy on this matter.

Various legal-economic mechanisms for the control of residuals into the environment are available to government. Basically these are:

1. Effluent standards to meet an ambient standard.
2. Effluent charges designed to meet an ambient standard.
3. Discharge rights designed to meet an ambient standard (Rights transferable).
4. Regulations on inputs to production.
5. Taxes on inputs to production.
6. Subsidies (Preferential tax treatment or granting of funds from one jurisdiction to another jurisdiction or to industry to foster reduction in waste discharges).

Perceptions of the objectives of a pollution control programme differ:[20]

—Environmentalists and industry people tend to assume that there is a single objective, namely, to preserve or enhance environmental quality.

—Economists view the objective in a somewhat complex fashion, namely that public policy is concerned with making the "best use" of resources and this involves not only proper use of the air and water but also making the user of the air and water take into account any benefits that are foregone by using the environment for waste discharge purposes. To the economist these are costs to be weighed so as to assure efficient use of all resources that enter into production. Economists usually view economic incentives as a means of:

(1) speeding up action to preserve and enhance environmental quality and

(2) fostering efficient uses of capital, labour and natural resources.

—The industry has been concerned at times that the real objective of "economic incentives" might be to use them as a means of producing revenues.

Probably the simplest and most palatable step in the direction of economic incentives that could be taken would be to adopt a system of charges that would apply:

1. During the time the waste discharger was making the adjustments to meet the effluent standards allowed by his permit.
2. If and when the waste discharger exceeds the discharges allowed by his permit.

The effluent charges should essentially be seen as administrative devices to ensure that the polluter pays, that the greatest polluter pays the most, and that industry is given every incentive to find ways to reduce pollution as well as to treat it.

For air pollution control there is legislation by the federal government as well as some provincial governments. Negotiations between the federal government and the provinces are taking place so as to establish basic national standards of emission that are acceptable to all. The national policy is based on the control of new stationary sources and the control of existing stationary sources within a negotiated period of time, using the best practicable technology. This 'control' is predicated upon 'containment-at-source'. High stacks (dilution and dispersal) are regarded as stop-gap measures while more suitable compliance plans are carried out.

In 1971, the Federal government enacted the Clean Air Act. In it there are provisions for agreements with the provinces for control programmes so that in certain situations the federal government may take direct action. This Act is also designed to set national emission standards especially in those areas where international agreements on air pollution are involved; to assist in developing uniform regulations across Canada and generally to prevent "pollution havens". The National Air Quality Objectives are listed in Table 7.

National guidelines for particulate emissions and water borne products are at present the subject of government/industry consultation. The air pollution problems associated with non-ferrous smelting and asbestos mining and processing are receiving particular attention. Efforts are underway to develop national guidelines for particulate emissions of SO_2. Specific SO_2 guidelines for Arctic mining operations are also expected shortly. Work will probably begin later this year on the development of national standards for arsenic emissions from roasting and smelting activities. With respect to asbestos fibres, regulations limiting particulate levels to 2 fibres per cc will be established this year to be in effect by 1978.

Canada is not alone in striving to control air pollution but some observers believe that conditions in many developing countries permit more liberal effluent and emission standards than can be tolerated in the more highly industrial countries. The policy maker should remember not to foreclose potentially important future resource options.

Table 7.

Contaminant	National Air Quality Objectives in mass units $\mu g/m^3$	
	Max. Desirable	Max. Acceptable
Sulphur Dioxide		
1 hour	450	900
24 hours	150	300
1 year	30	60
Particulates		
24 hours	—	120
1 year (geom. mean.)	60	70
Carbon Monoxide, $\mu g/m^3$		
1 hour	15	35
8 hours	6	15
Total Oxidants		
1 hour	100	160
24 hours	30	50
1 year		30

Coal-fired electric power plants, steel makers and smelters are the principle source of industrial air pollution in Canada. There are 16 non-ferrous smelters in Canada: 6 copper, 4 zinc, 2 lead and 4 nickel. Two of the copper smelters, 3 zinc, 1 lead and 2 nickel smelters are operating at various levels of SO_2 containment. The one hydrometallurgical smelter (Sherritt-Gordon) has 100% SO_2 containment.

Two nickel smelters and one copper smelter are under government orders to increase SO_2 containment to 80–86% by 1978.

The problem of sulphur disposal in the smelting process of non-ferrous metal ores is fundamental in Canada. It stems from the fact that our ores are essentially occurring in the form of sulphides. For example, copper content in concentrates of Western Canada tends to be 25–30% while sulphur is 25–40% and iron 20–25%. In smelting, the iron comes off in the slag and the sulphur is usually carried into the waste gases. A flash smelter producing 150,000 tons of copper annually, can produce approximately 500,000 tons of sulphuric acid per year. Past practice in most cases has been to vent the SO_2 into the atmosphere with the waste gases. This practice continues in the developing countries and is still evident in North America. Nevertheless present discussion concerning the development of new copper smelters in Western Canada indicates that the smelters will have to capture a minimum of 90% of all sulphur produced in the operations. It must be recognised however that attainment of these environmental standards will impose on smelters additional operating and capital costs. These could be offset to some degree if a market is found for the sulphur products.

Generally speaking in cases of smelters already in operation, the choices open to achieve further control of SO_2 are:

1. to modify the process
2. to limit production generally or according to weather conditions
3. to install equipment to meet emission standards
4. to implement a compliance programme over a specific time or
5. to increase efficiency of plant operation.

Certainly, these choices are not exclusive and the smelter operation can function with more than one.

With respect to water quality a number of provincial statutes are related to controls on water quality and effluent discharges. However, the most important piece of legislation presently in force is the Canada Water Act because it forms the standards to which other laws conform. There is yet other federal legislation, such as the Fisheries Act, that has a profound impact on water quality control. Special provisions for the North and the Arctic embodied in the Northern Inland Water Act give the Crown all rights to both surface and underground water in the Yukon and the Northwest Territories. In the North there are also Territorial Land Use Regulations which apply to specific regions because of their unique resource characteristics, and the need for environmental preservation.

At the Federal level efforts are also underway through the establishment of government/industry Task Forces to develop national guidelines for reducing the toxicity of effluents. The proposed maximum levels for a number of pollutants are presented in Table 8.

Table 8.

Parameter	Concentration
Arsenic	0.5 mg/1 total
Copper	0.3 mg/1 total
Lead	0.2 mg/1 total
Nickel	0.5 mg/1 total
Zinc	0.5 mg/1 total
Suspended Solids	25 mg/1
Radium 226	10.0 pCi/1 dissolved
pH	6.0 minimum

In land reclamation all four provinces which have surface coal mines; namely British Columbia, Saskatchewan, Alberta and New Brunswick have legislation governing reclamation as well as surface mining practices. The concept of land reclamation has now obtained broad acceptance throughout the industry, and some of the more recent reclamation proposals related to new open-pit base metal mining are extensive, and come as part of the overall feasibility and planning studies associated with new mineral development. As of last year all the coal companies have a reclamation programme as part of their overall mining plan. The costs of reclamation have been found to vary considerably depending on which type is carried out. Average cost per acre figures provide only marginally accurate estimates at best. Revegetation alone may cost only $100–500 per acre, whereas full reclamation in a mountain region may cost $1,500–4,000 per acre. Full reclamation although uncommon in North America, has been applied extensively in the U.K. and West Germany.

Industry is increasingly coming to accept their share of responsibility for both cleaning up past environmental pollution and preventing additional harm. To assist industry in offsetting these costs, government has introduced an Accelerated Capital Cost Allowance Programme and has granted exemption from federal sales tax for equipment designed to control pollution resulting from a production process. Funds can also be made available to industry for the development and demonstration of new pollution abatement technology.

To sum up the current situation in Canada, one might say the art of environmental management is really only beginning. Beyond the deleterious aspects of pollution clean-up and anti-pollution measures, the subject of mineral development has to be examined more closely but from a multi-disciplinary point of view. The total environmental setting of a project must be considered, that is to say, the effects on human health, the landscape, the biological base, possible recreational needs and water mangement, as well as aspects of transportation communication, and hydro-electric power.

Conclusions

In the paper a broad outline of the mineral development and the environment interface was presented. The importance of mineral development in economic growth was illustrated with specific reference to Canada. Also the nature of potential conflicts and costs between mineral development and the environment were highlighted. Finally, some policy options currently in effect in Canada designed to harmonise resource development and conservation objectives were reviewed. It remains to summarise the main conclusions that follow from the above analysis:

1. Government's role must be to maintain a balance between development and environmental protection. This balance must take into consideration a variety of interests from local and immediate to national and long range. Environmental problems have a cumulative impact and if remedial action is taken only at later stages of a country's development, the costs will

be considerably higher. The government's task is to define the framework for environmental protection within which industry must operate, but not necessarily to tell industry specifically how to meet these requirements.

2. Basically, safe thresholds do not exist. In terms of policy, this means there are no magic numbers of neat value-free criteria; all environmental standards will require political judgements and careful attention will have to be given to facts and analytical methodologies before acceptable and effective standards can be establised.

3. In reference to containment of pollution at its source by use of the "Best Practicable Technology": although technology has always come up with answers in the past, some solutions have led to other social, environmental or economic costs. Today, these costs are regarded more seriously than ever before; in fact, they are setting a practical limit in many instances to further mineral development. Lack of available technology is not the principal barrier to air and water pollution control. The basic technology is known for most pollution problems, even though engineering applications may be poorly developed. Pollution control is primarily an economic/political problem. It is often developed only with reluctance since it is seldom profitable.

One of the most significant contributions of the Stockholm Conference on the environment was to make clear that many of the most acute environmental problems of the developing world arise from poverty and inadequate development, and that the routes to a better quality of life lie not through seeing environment and development as separate and often contradictory notions, but through the integration of environment with population and resources as key and interacting elements within the framework of the development process itself.

4. In Canada, up to now, the problem of pollution control was measured in terms of dollar cost only; but as the Prime Minister, P.E. Trudeau said in January, "It is becoming clearer every day that we must become more sensitive to the true cost in terms of ill health, and long-lasting damage to nature and to the quality of our lives. We must begin right away to appreciate the growing seriousness of the problem, become more willing to accept our share of the responsibility, more receptive to the need for basic change in some old and familiar industrial and personal habits."[21]

5. Each country faces a different set of problems and values. Consequently, different countries will make different choices between environmental and other goods. There is, however, a need for reconciling the interests of government, industry and the consumer everywhere. The interrelationships of national and international policy must be fully recognized at every level of policy formulation.

Acknowledgement The author would gratefully like to acknowledge the assistance and advice provided by D. Brooks, J. Desmarais, R. Woodbridge and various other colleagues from the Department of Energy, Mines and Resources in the preparation of this paper.

Discussion Summary

In discussion it was pointed out that the industry has been driven to short term solutions but that pollution costs are long-term. Delay in implementing controls may lead to much higher costs. Environmental demands can lead to closure of works but the resulting unemployment may lead to political difficulties and the new industry of environmental management will not absorb many of those displaced from the mining industry. In the United States, for example, the recovery of sulphur from effluents of smelters creates additional jobs and similarly, control of asbestos milling does likewise but twenty or more smelters will close because of non-compliance with environmental protection laws. In Canada at Sudbury compromise solutions to avoid wholesale unemployment and loss of production will be accepted. In tropical conditions too little is known about relevant environmental interactions but public pressure had stopped copper smelter development in Puerto Rico and the acceptable amount of suspended matter in waters has been complained of by farmers in the Philippines. There are no world environmental standards but the World Bank normally requires that USA standards be met. With regard to control of sulphur dioxide emissions the long-term answer is concentration and containment, the short term solution is dilution and dispersion from ultra-high stacks. New smelters can be designed for 90% containment of sulphur dioxide.

1. David B. Brooks and P.W. Andrews, "Mineral Resources, Economic Growth and World Population", *Science*, Vol. 185, No. 4145, July 5, 1974.
2. Energy, Mines and Resources, "Towards a Mineral Policy for Canada, Opportunities for Choice", Ottawa 1974.
3. Jeremy Bentham, *An Introduction to the Principles of Morals and Legislation*, 2 vols. (1783; rpt. London E. Wilson, 1823)
4. Environment Canada, Environmental Protection Service, "A Nationwide Inventory of Air Pollutant Emissions to 1970."
5. L.M. Whitby & T.C. Hutchinson, "Heavy Metal Pollution in the Sudbury Mining and Smelting Region of Canada. Soil Toxicity Tests". Published by Elsevier Sequoia S.A. for the Foundation for Environmental Conservation, 1974.
6. Report prepared by the Ontario Water Resources Commission.
7. A.J. de Villiers and P.M. Baker, *An Investigation of the Health Status of Inhabitants of Yellowknife, Northwest Territories*. Dept. of National Health and Welfare, Ottawa, 1969.
8. Henry Ford II, in an address before the annual convention of Sigma Delta Chi, professional journalistic society, Chicago, Illinois, November 13, 1970.
9. Neil W. Chamberlain, "*The limits of corporate responsibility*." New York: Basic Books Incorporated, Publishers, 1973.
10. M. Strong: Address to the Men's Canadian Club of Ottawa, February 12, 1975.
11. D. Brooks, "Conservation of Minerals and the Environment", (unpublished), 1975.
12. World Bank, *Environment and Development*, June 1975.
13. *The Concept of Environmental Costs*. Environment Canada, Policy, Planning and Evaluation Directorate, 1974.
14. J.H. Dales, *Pollution Property and Prices*. University of Toronto Press, 1968.

15. Dept. of Commerce and Environmental Protection Agency, *The Economic Impact of Pollution Control*, A Summary of Recent Studies Prepared for the Council on Environmental Quality, Washington, D.C., March 1972.
16. OECD, *Economic Implications of Pollution Control*, Paris 1974.
17. According to the 1975 estimates of the U.S. Council on Environmental Quality (CEQ) the US is expected to spent $194.8 billion during the 10 year period from 1973–1982. This is the staggering cost of environmental legislation. The average cost per person in the US which has been $35–40, will more than double in 1976.
18. Best Practicable Technology is a concept used in both the U.S. and Canada and really means "Best available technology". The application of this concept implies an essentially clean effluent in all respects. On the other hand, the phrase "Best practicable" implies some level of treatment less than the best available and is generally taken to be the level that can be borne by industry without intolerable economic consequences.
19. Richard Zerbe Jr., *The Economics of Air Pollution: A Cost Benefit Approach*, Ottawa, 1974.
20. See papers presented at the Conference on Use of Economic Incentives for Air and Water Pollution Control. University of Victoria. June 12–14, 1974.
21. P. E. Trudeau: Prime Minister's Address on the Economy, Ottawa, January 1976.

PAPER 10

FINANCING MINERAL DEVELOPMENTS

Alternatives available to Developing countries

by Peter A. Glenshaw

THE PURPOSE OF this paper is to give a brief overview of how mineral development projects can be financed in developing countries. It discusses the financial needs of projects currently being developed all over the world, the requirements for arranging financing and the public sources of finance of interest to the mineral processing industries of the developing countries, as well as the whys and wherefores of project (as opposed to the more traditional corporate) finance.

The paper tends to use the idiom of the Western financial world with which the writer is most familiar and refers to a fundamental principle; namely, that in order to launch a successful project, one must be able to demonstrate clearly that the value of its output will exceed the cost of launching and sustaining it. A corollary of this principle is that the project with the greatest excess of value of output over costs will tend to be chosen first. Whatever measure one places on the value of output and costs these principles will apply.

The paper is not intended to be a "cookbook" with magic recipes on how successfully to obtain funds in a world increasingly short of capital. Rather, it tries to explain the essential requirements for successful financing, and to show what the providers of finance have to look for in order to remain in business. In this way, it can be said to provide a useful checklist of points to be met, if one is to have any hope of competing for the very substantial funds required.

The Financial Needs of Mineral Development Projects

The annual listings of mineral projects give a good indication of how diverse these needs are. Some 600 projects covering nearly 30 minerals worldwide are listed in the *Engineering and Mining Journal* of January 1975. The investment costs for 375 of these projects are given, the average of which is $115 million. Most expensive on average are the aluminium projects at over $230 million each, followed by iron ore ($195 million), nickel ($160 million) and copper ($153 million). At the bottom of the scale are projects for the light metals—titanium, lithium and magnesium—projects which average about $16 million each. The least costly project listed is a $1.5 million expansion of salt production in Canada, while the most costly is the 80,000 ton per year aluminium complex of Aluminio do Brasil whose cost is put at $2.5 billion.

The cost of an individual project will depend on many factors, of which size, complexity and type of mineral are probably the most important. However, as mineral developments are increasingly based on low grade ores in more remote areas, infrastructure type costs become as significant. As an illustration it is interesting to compare copper mine/concentrator projects. The following table lists such projects in varying stages of development, and shows the diversity of sizes as well as cost per annual ton copper.

Table 1: Copper Mine/Concentrator Projects

Project	*Location*	*Size '000 tons yr of Cu**	*Investment Cost Total $ million*	*Investment Cost Per Annnal Ton Cu* $*
Minero Peru	Michiquillay Peru	120	700	5833
Cities Service	Pinto Valley California	65	100	1538
Anamax	Twin Buttes Arizona	57	93	1635
Phelps Dodge	Metcalf Arizona	50	180	3600
Atlas Mines	Cebu Philippines	50	80	1600
Highmont Mining	B.C. Canada	34	70	2058
Cia Minera de Cananea	Mexico	29	75	2586
Lepanto Mines	Negros-Occidental Philippines	26	68	2615
Hindustan Copper	Madhya Pradesh India	13	70	5384
Mitsui Omnium	Morocco	10	16	1600
Black Sea Copper	Turkey	10	23	2300

Source: *Engineering and Mining Journal*, January 1975.
* Cu as contained in the concentrate.

Because of diversity one cannot generalise on costs. Each project has to be considered on its own merits. Even the usual expectation of economy of size does not seem to hold when the above list of projects is examined. It is likely, however, if each of the projects was put onto exactly the same basis; particularly by removing costs for infrastructure and other items unique to each project, that the larger projects would show lower unit investment costs than the smaller projects.

Very small projects must suffer from the burden of having high infrastructure and other development costs associated with them. They are likely to be uneconomical for these reasons. If there are more than one such small projects there are benefits in forming cooperatives of mines to enjoy the economies of scale. When such cooperatives also include a banking or other financial element, they are made even stronger. Some of the most successful mining "houses" in the world started from the consolidation of numerous small mines and financial institutions, and many still carry the term "Consolidated" in their names.

Another benefit of consolidation is that as a project becomes more viable, it is more likely to be able to engage consultants and other personnel skilled in modern techniques of project identification and development. In this way communication can be effectively established with outside agencies essential for the project's success. Agencies that can supply finance, technology and markets to the project are staffed with such skilled personnel, and employ techniques that are themselves always being enhanced and developed. The inability of these agencies to communicate with projects in developing countries has often resulted in the projects remaining dormant, even though they might be basically technically and economically sound.

The Key Requirements for Project Financing

At issue in every instance of financing is the concern whether the money extended by the financier will be returned by the user—preferably with interest—in the not too distant future. Three key requirements can be distinguished that should be met to ensure that this basic issue will be satisfied:

The project should be based on a sound techno-economic concept;
The project should be credit-worthy; and
The project should have some form of financial guarantee.

Sound Techno-Economic Concept

The project promoter must be able to demonstrate that the project can technically perform what is expected of it. It should not be dependent on untried technical methods unless there is clear evidence of substantial benefits from the new methods and of superior technical support to see the prototype technology through its inevitable testing and trial phase. There should be a fall-back position where alternative—perhaps nominally more costly—but proven technology can be substituted if the prototype proves difficult to get going. This requirement applies to the subsidiary parts of the project as well as to the main part. An untried transportation link can be as ruinous as untried basic process technology.

Nothing is more comforting to a project financier than the "repeat" project where every element is a copy of a previous successful project, especially the technology employed.

Besides proven technology, the project must be based on sound economics. It is no use trying to promote a project making blue suits when the market clearly wants nothing but green suits. Market demand is the key economic element and the usual stimulus to project identification. Availability—either through ownership or through supply contracts—of the raw materials is equally important. In both instances the prices of supply and demand determine if the project will be a paying proposition. In large capital intensive projects, the demand price is more important than the supply price, as capital costs are usually as important as other input costs.

The cost of capital can be difficult to put a figure on, and normally a number of levels of cost will be tested for a series of different financial plans. A basic need of a financial plan is to ensure that senior obligations will be met, such as the servicing of debt and other claims, such as taxes. Then comes the return to the owners who may be satisfied merely that the project operates on a cash breakeven basis as they get their substantial return from activities connected with the project—such as using the output or supplying the raw material. On the other hand, they may demand a cash return as dividends.

Usually, the cost of capital is not figured with any degree of accuracy until it is necessary to decide between possible alternatives at a very mature stage of the project. Until then, a general cost level for the industry, or for projects of a similar nature, is used to estimate the minimum cutoff point below which the rate of return should not fall.

Credit Worthiness

Formal systems of measuring the degree of credit worthiness, or the credit rating of a user of the funds, have become more popular as the relationship between the providers and users of funds has become more institutionalised and less personal.

The techniques of examining a person's credit rating are well established, whether the person be an individual or a corporation. The main criterion has been the past record of financial dealing. When that record has not been available because the person is getting into a form of business new to him, credit has not been easily or lightly given. A stricter mathematical analysis of the person's prospects in the new business is required, the credit agreement is tightly binding and well secured, the funds are expected to be returned as soon as possible, and generous cash cover of the debt service is required.

The same techniques are being applied to projects based on the concept that a hypothetical corporation is in existence for the sole purpose of implementing and operating the project. Often there is a real corporation set up for this purpose. The point is that hypothetical or real, the corporation does not have a credit record, but by careful analysis and structuring the credit risk can be made acceptable, so that funds can be extended for the project.

Financial Guarantees

The question of guarantee is another key to financing. If the deal does not work out, is there someone who will repay the financier, at least in part?

In the case of personal or corporate financing, the idea of a guarantee is implicit in the limitation that the financier will impose on the amount of money extended. The deal will be limited enough so as not to involve all the resources of the user, so that if it goes sour, he will have something left over, either to start again and then pay back what he owes, or to pay back directly.

In the case of a project this implicit reserve is not necessarily available unless the project sponsors have existing banking relationships with the financiers which they will want to keep intact. Even in such cases, formal guarantees may be asked of the sponsors. Often though, the sponsors will not have the wherewithal to stand behind such a guarantee, but the project itself may have assets which could be pledged, or may have contracts that are valuable in themselves, and that can be pledged to back up the repayment guarantees. If a long-term sales contract for the output of the project is envisaged, it is often the most valuable tangible asset (in its broader sense) that could be pledged as a financial guarantee.

The Project Sponsors

There are four main groups of what can be called sponsors that have to be involved in a project to make it successful. They are:

the project promoters who are often its owners as well
the project's suppliers
those who purchase the project's output, and
the project's financiers.

The Project Promotors

Generally when one speaks of the project sponsors, one refers to the people—individuals or corporations—who conceive the idea of the project and go about drumming up interest in it until it becomes a workable proposition. Another term for this activity is project promotion, and the sponsors are referred to as the promotors.

Whatever they are called, the people who put the project together are the key to its success or failure. The project concept depends on the existence of people with a driving force strong enough and with enough stamina to see a project through its most difficult period—from which it starts as a gleam in someone's eye to the point when sufficient interest has been generated in it so that it becomes self-sustaining. The fact that this process may last a decade or more speaks for the kind of strength and staying power that is needed. For this reason it is not surprising that government entities—whether local or national—are often the main promoters of projects in developing areas.

It is the large corporations that are the main promoters in the areas where industrial development is well advanced. They used to have a large role in promoting industrial and mineral projects in developing areas until recently, but increasing local desire to hang onto resources and not be at the beck and call of the large outsider has tended to put them off. Again their position has been hurt by well publicised accounts of out and out grabs, which is a pity, as so much of the expertise and strength needed for the exhausting business of project promotion resides in the large corporations.

As the project gathers momentum, more parties will join it and as they become committed, they could be also termed project sponsors. Their interests will vary considerably, and the role of the original sponsor, or promoter, then shifts from that of garnering interest to that of matching the often conflicting demands of the many groups which have signed on to the project. It is here that the role of project finance becomes of key importance, as the norms required by the financial "sponsors" become the norms of the project and can be used to resolve differences. Thus pressure may be put on say the buyers of the project's output who might have been holding out for a price of 80 when a fair market price might have been 100 and the financiers show that they need 90 to give some assurance of debt service cover. Even before the financiers actually join up, a promoter with sufficient knowledge of what they will be demanding is in a good position to modulate the demands of the various parties.

An aspect of promotion vitally important to the success of project implementation is that of enlisting governmental or political support. No matter how good a project may be technically and economically, if it does not have the support of the leaders of the political regimen in which it is being promoted, it is doomed to failure. Politics, even in its broadest sense, being the fickle force it is, is often the root cause for success or failure in getting a project implemented. The project record is full of cases where clear losers were pushed ahead and where clear winners still wait to be implemented. But, whereas the cynic may say that the political driving force is the all or nothing of project success, there is also ample proof that properly conceived projects, professionally put together, will sooner or later succeed. The garnering and maintaining of political support for the project is one of the many skills required by the promoter.

The Project Suppliers

These are the entities which will provide the know-how, the equipment, construction services, the raw materials (although often the owner is the supplier of the raw materials) and the other operating supplies. Their early involvement is important, so that accurate cost estimates can be made for both the capital and operating costs. At the earliest stages the suppliers will give estimates with an accuracy useful only for budget purposes. These general-accuracy estimates help the promoter sort out which of the many potential suppliers best suits his purpose, and he then makes firmer and firmer arrangements with them until a contractual type of commitment is made.

Normally, the promoter will wait until the next two groups are engaged before finalising contracts, so as not to preempt any beneficial arrangements for the project. For example, if he could get bridging finance from the financial group, he might be able to make substantial down payments on equipment and get a discount on the price. He would, therefore, hold off making closing equipment contracts until he could see how his financing was going to turn out—unless, on the other hand, a long delay for this purpose might lose him an opportune moment in the equipment market.

The Purchasers of the Project's Output

If the output is being sold in a generalised market, no contractual situation will be involved, but careful research will be needed into the absorptive capacity of the market in terms of quantity and the competitive position of the project compared with other suppliers to the market. Most important of all is the price expectations for the product. It is necessary to convince the financiers that the project will sell its output at a high enough price to give it an adequate income to provide a minimum return on the investment.

The price at which the production is sold is the key to the financial viability of a project, as one can generally expect that, except for wildly speculative products, the quantitative output can be disposed of—if only at distress prices. Price estimates in a generalised market are extremely difficult in these times of galloping inflation and chronic swings from over to under supply in basic commodities. Even in a situation where the purchasing group is clearly identified and able to be brought into a contractual position, it may be difficult to get any firm pricing arrangements. A compromise in such cases is to have the purchasers agree to a minimum price which will provide the minimum debt service or other cover that the financiers will require. When a sales contract is envisaged, it is often the basis for the financial arrangements and, therefore, must be very carefully drawn so as to satisfy the financiers but not overly restrict the project—particularly in the non-normal times of boom and recession in the business of the project.

The Project Financiers

The last group normally to be brought in is the financial group. Of course it is likely that considerable finance or promises of finance will have been gathered in the process of engaging the suppliers and purchasers. Suppliers of equipment and services can usually be expected to provide or to offer suppliers' credit of one type or another. The purchasers may be asked to show good faith by putting up some cash in the form of equity in the company. Other suppliers may put in their supply without payment as a form of equity supplied in kind. The owner/promoter will have sunk some costs which can be regarded also as equity, and will be prepared to put in more equity in the form of cash.

The financiers will often then be asked only to close the financial gap. As this is usually the most difficult part of the financing to provide, there are often delays and holdups while the financial group tries to restructure the financial pie to add some sweeteners to the unpleasant part it is asked to swallow. The gap often is concerned with the least secured, longest repayment parts of the project's financial plan, and it is not surprising that the financial group will look for some restructuring. For this reason it is well not to leave the financial group out of the picture for too long—or alternatively to have someone in the promoter group who is familiar with the financiers' needs and wishes.

Relationship of the Four Sponsors

While it is possible to distinguish the four major groups involved in putting a project together, in practice the roles are often mixed and the dividing lines blurred as the groups are tightly interrelated. It goes without saying that the more clearly the interrelationships are defined, the clearer will be the syngergism of the project and the greater will be the cohesive forces driving the various groups toward successful implementation of the project. Unfortunately, the opposite is also equally true. The most difficult projects to launch are those in which the various participants have only a loose relationship to one another and to the project. The art of developing such relationships to overcome doubt, reluctance and ignorance is the mark of the successful project promoter, and the ability of the relationships to survive times of stress and difficulty is the mark of a successful project.

Factors Outside the Project

Once the key requirements of the project have been met, and the four groups of sponsors have been identified with some degree of certainty, the project promotor will want to ensure that the matters outside the project boundary are in hand. He will be particularly concerned to ensure that:

there will be an adequate supply of skilled personnel at all levels and that

there will be adequate infrastructure to ensure the operation of the project.

Supply of Skilled Personnel

A successful project will have adequate resources of adequately skilled personnel at all levels. The assurance of experienced and trained management at all stages of the project, but especially through the critical construction and start-up phases is one of the most comforting that all the different project sponsors could ask for. The more complex the project, the more important is the quality of management required to see that it is successfully completed.

Yet it is surprising how often the owners will employ second-rate management supposedly in the name of cost savings.

It is true that top quality management and technicians are a scarce commodity in the world today and it is not surprising that their salaries are high—often many times higher than those of the owners or other sponsors. Yet the overall cost of management in a large capital intensive project is minute compared with the costs of other items, or with the cost of error or delay that can be caused by incompetent personnel.

As important as good management, a successful project needs a well trained and stable work force. Training facilities must sometimes be included in the project if the skilled labour supply is limited. They will be needed anyway to keep on upgrading the skill level of the workers. The social environment should be conducive to encouraging stability, as the cost of rehiring and retraining a too mobile work force can prove to be ruinous. The remote area project does well not to underestimate the costs of providing adequate training and social environment, even though they do not directly contribute to production. The damaging effect that untrained and disgruntled workers can have on productivity and—literally—on expensive equipment can be many times more costly than the provision of adequate training and social amenities.

The provision of what from the project's point of view are basic necessities can often put the project labour force well ahead of the local average citizen and psycho-political difficulties can be encountered. If the project's foremen are going to live in better houses and have access to a better club than the government can provide for say the magistrate, there is bound to be friction. Salary levels can be even more contentious.

There are no easy solutions to these problems, but it is generally better to try to use the project to upgrade the general levels rather than risk getting inferior personnel through it coming down to the general level.

Adequate Infrastructure

Electric power is needed and a public utility is technically and economically able to supply power—what contract for power supply will the financiers require? What contract will the public utility be able to commit itself to? A new road connection is needed. Will the governmental road entity build it, or should the project build it and then turn it over to the government? How will costs be met in the latter operation? Can the project use long-distance radio communication until the government or public utility systems are installed? In a remote area, who will provide housing and other social amenities? In a partly developed area, where the project would have an immense social impact such a question is often more difficult to answer as entrenched interests have to be met. Are trans-ocean shipments involved? Are special vessels required, or will the project rely on the world liner/tramp market? What state will it be in when the project needs it? Taxation rulings, concession rights, corporate registration procedures, and other legal arrangements must be tied up.

The arrangements outside the project to ensure that adequate infrastructure and legal matters are in hand can often take as long and be as complex as the arrangements for the project itself. Where considerable new infrastructure is required, such as in remote areas, another difficulty is that control of seeing that the project's needs are properly implemented slips out of the owner/promoter's hands, and they become dependent on governmental entities with their own set of priorities which are rarely the same as those of the project. Even more difficult is the situation which often occurs when government agencies see the establishment of a major new mineral or industrial venture as the nucleus of a development pole in an underdeveloped area, and try to accomplish years of pentup development attempts through the project. A pendulum effect is often noticed in which the government agencies move from dull reluctance, if not outright opposition, to overenthusiastic support and start overloading the project with elements it cannot possibly bear—at least not in its initial phases.

Completing the Financing

Once the promoter has ensured that the other "sponsors"—suppliers, purchasers and financiers—are ready to commit themselves and that all are convinced that the project's basic elements are intact, i.e., based on a sound techno-economic concept, that the non-project arrangements are in hand, and that it is assured of good personnel, he is ready to tie the knot, close the deal and complete the financing. Whatever the precise course of events, a moment will come when major commitments must be made and the more important arrangements must be completed, particularly those which will assure the project that funds will be available.

Whatever the dynamics of closing the deal, a financial plan will have to be agreed on and accepted by all groups. The split between debt and equity will normally get the most attention. Rules of thumb such as that there should be at least 40% equity, or that debt will be used only for foreign exchange expenses—or only for equipment purchases, will be the initial guides. A second stage is the attempt to allocate specific sources of funds to specific expenses.

Loans can usually be found to finance the purchase of land, equipment, and services, while equity is usually called upon to cover the pre-operating expenses, working capital and financing costs including interest during construction. But there are no hard and fast rules and projects have been known to be launched with practically all debt or all equity. It should be pointed out that an all-debt project can be put together only if all arrangements are completely secure and usually requires that the sponsors give watertight financial guarantees to the lenders.

The other main element of concern in the financial plan is whether it will adequately cover the needed investment costs. Today more than ever, cost overruns of serious proportions are rampant. Arrangements to finance overruns, or—as it is sometimes called—to complete the project, are of major concern, especially to the financial sponsors. Sophisticated methods of estimating and allowing for inflation during construction are being devised but with no assurance of being successful. Lenders will often require the equity partners to make legal commitments to supply the overrun financing and/or to complete the project. Completion is often defined as being well after construction has finished and a usual requirement is that the expected levels of production be convincingly demonstrated before the project is said to be complete.

Once the financial plan is drawn up, i.e., once sources have been found and have committed themselves to cover all the expected investment costs, the project financing can be said to be completed. The final stages of actual financial deal-making are thus seen to be only the finishing touches of a long and drawn-out process—much of which has little to do with finance as such.

Sources of Finance for Mineral Industry Development

Sources of finance can be divided into two main categories—the private and the public sources. Undoubtedly, the private sources are still the main contributors to the development of the mineral industry on a world-wide basis, and probably also represent the majority of the existing investment on both a world-wide basis and in the developing countries. However, these are times of reluctance on the part of private investors to invest in foreign countries and on the part of developing country governments to receive such investments for natural resource development. It is, therefore, likely that the flow of new investment funds to developing countries for mineral processing is now dominated more by what can be called public funds.

Such public funds would consist of appropriations from national budgets, loans and/or equity supplied or at least backed by governments, and foreign funds from such sources as the international development banks, national aid agencies, export credit organisations, the newly created petro-funds, and the UN system. This section of the paper deals with the chief characteristics and requirements of these public sources.

Government Supplied Funds

Generally funds from the national budget are available for infrastructure, or for the purchase of securities in holding companies which the government will use as its vehicles to manage its interests in mineral processing industries. The holding company vehicle is a useful device as it frees the operation of the industries from the structures of government regulations—and also limits the liability of the government in those operations. The ability of a government to provide funds to either the holding company or for infrastructure will, of course, depend on its financial condition, and the other demands on the budget.

However, the fact that the funds will hopefully result in the production of a revenue, and of increased taxes, should give the industry a somewhat easier time with the budget managers, than its non-revenue-earning competitors. Often budget funds can serve a useful purpose as bridging finance, keeping the project going, and building up the confidence of other investors. Government funds are most appropriately used for local expenditures, which are often very difficult to finance from foreign sources.

Where the government does not want to commit its own cash resources and has a healthy financial situation, it can stand as surety behind share or bond issues for a mineral processing development—probably through the holding company intermediary. Such government guarantees will probably be asked in any case by foreign sources of funds—and there is no reason not to use the same device to try to raise local capital. The government has the advantage of being able to offer tax-free interest or dividends on such backed issues. It is likely that locally issued securities will have to take a junior position to any foreign investments in the project.

International Development Banks

The World Bank Group, and the Inter-American, Asian and African Development Banks are the leaders in this group. Financially, they operate as true banks—making loans and bringing co-financiers into projects. They are not equity investors (except the IFC) and generally demand substantial equity investment by others in the industrial projects they finance. However, they go well beyond the services provided by commercial banks, and make major contributions to the development, identification and preparation of their projects.

The terms of the loans will vary from bank to bank, and from project to project, but in general the ultimate borrower will be expected to pay something like a commercial rate of interest for a 10 to 15 year loan which will carry a grace period equivalent to the construction and start-up time. The banks may well lend their funds initially to the host government often at very concessionary rates, but they will invariably demand that the host government on-lend the funds under commercial conditions. The reasoning behind these demands is that the international development banks which depend on developed nations for most of their funds, cannot appear to offer subsidised financing to industries which will sooner or later compete with others in the developed countries.

The four major international development banks can generally deal economically only with projects costing into the tens of millions of dollars because of their high cost of processing each project—and their demand for rather elaborate procedures in the use of the funds they extend.

Smaller projects will, therefore, find the regional investment and development banks more suited to their needs. Such institutions are, for example, the Caribbean Development Bank which serves a limited number of countries—and the local development bank which serves only the nation in which it is located, but which usually has foreign shareholders—including the four major international development banks. Through these smaller banks, access to international funds and some of the expertise of the majors can be obtained, without the sometimes crushing weight of their elaborate systems.

National Aid Agencies

The national aid agencies of the developed countries were a most important source of investment capital for developing countries in the past. Today, their former dominance of the flow of development capital is being taken over by the development banks, and their role is possibly even more politically inspired than it was formerly. To the extent that the recipient nation is favoured by aid donors, the national aid agencies remain an important source of funds.

National aid funds tend to be earmarked for social programmes of immediate political impact, rather than for industrial projects with long gestation periods. On the other hand, such funds may be easily obtainable for the development of natural resources of interest to the donor country. The terms attached to such funds vary from outright grants to semi-commercial bank terms—with probably the bulk having a very long term (20 plus years) repayment obligation at a nominal service fee.

Export Credit Institutions

Most of the developed nations have export credit organisations of one type or another. On one end of the scale are government owned bank-like institutions which advance cash—usually in the form of loans—to foreign purchasers against the supply of goods or services from local firms. At the other end are institutions which guarantee payment to local firms for goods and services they deliver to foreign buyers. The terms of the loans and the guarantees again vary from case to case, but tend to follow those set in the financial markets—but often are slightly better.

Export credit of whatever form is available only to those buyers in nations where the credit-extending nation seeks to obtain some political advantage, or to cement an established political position. It is limited too in that it can only be used to obtain goods and services in the credit-extending nation, when better prices could perhaps be obtained elsewhere—especially through international competitive bidding. There is a clear tendency for suppliers to increase prices when they know they are bidding in the limited—and often non-existent—competition in a single nation. This is especially true with regard to the specialised goods and services needed for large, modern, mineral processing plants.

Nevertheless, export credit can form an important part of a financing scheme, and has the advantage of being extended with less elaborate investigations and procedures than with the international development banks. When the goods and services can be "packaged" in rather large bundles, it is also possible to obtain a form of international competition by seeking both the goods and services and the financing in each offer. When supplier nations are short of orders such tactics will be quite effective—and it is often possible to get a good proportion of the local supply also covered by including it in the bidding package.

The Petro-funds

The so-called petro-funds are included here as a special category only because of their topical interest. In reality, they combine the characteristics of international development banks, aid agencies and export credit organisations. The dominant characteristics are determined by ownership by one nation or by more than one. In the latter case, they tend to act more like international development banks—although with a much more restricted purpose, a more limited geographical spread of clientele and a more clearly political motivation than is the case with the four major international development banks. Those funds operated by a single nation will be even more political in motivation, but unlike the purely export credit institutions, will often make concessionary financing available.

Many of the funds in the large petro-countries have had to limit their foreign operations recently in order to meet the needs of massive internal development programmes, whose cost has increased both due to inflation, and to somewhat over-ambitious planning. Another limit to the operation of these funds—both at home and abroad—is the lack of adequately trained staff to carry out the minimal amount of investigation and preparation needed to ensure that they would make sound investments. Again, these funds will be a very useful source to countries possessing minerals desired by the nation operating the fund. If the mineral-rich country can also demonstrate the ability to process the minerals locally, there should be no reason why such funds could not be used in mineral processing projects.

The UN System

Apart from the major international development banks which often carry the status of specialised agencies of the UN, the other sources of funds from the UN system of interest to the mineral processing industry are the UNDP and the newly created United Nations Revolving Fund for Natural Resources Exploration, which is administered by the UNDP. It must be pointed out that the Revolving Fund is not available for direct investments in the mineral processing industries. On the other hand, if it can free funds that an integrated mining and processing company would otherwise dedicate to exploration—it can be said to assist indirectly in investment in all aspects of mineral development.

An important difference between the Revolving Fund and the usual UNDP funding, is that there is a repayment obligation to the Fund out of the proceeds of successful operations which it initiates. The resources of the fund are not expected to be large but it does represent a new departure in the UN system, which if successful, could become an important source of finance in the future.

The UNDP too is limited in the amounts of funds that it can supply. The competition for these funds is very strong as they are often the only sources of finance available to developing nations—especially for social programmes which will not generate direct revenue. Despite these limitations, when the fund is in a position to carry out feasibility studies, it can serve a very useful purpose in the initial stages of a mineral processing project, by enabling the project promoter to obtain highly professional services for the preparation of the feasibility and other studies that will be his best arguments in trying to extract the much larger investment funds from reluctant lenders and investors.

Discussion Summary

In discussion, interest was expressed on the time-scale for development which was stated to be a decade though it was pointed out that many potential projects have been known for several decades. Only in the most favourable conditions could shorter times be envisaged. Developing countries can provide an input of the mineral, the land, labour and law but often must rely on finance from developed countries which in turn relies on techno-economic guarantees. This may limit available techniques and also limit choice of supplies of equipment, possibly supplied at inflated prices.

Decisions on viability may also be affected by positions taken by large producers and consumers which have an influence on sources of finance. The promoter clearly requires experience and skill to handle the evaluation without bringing in too many interested parties at an early stage and by relying on financiers' known reputations rather than sales efforts of suppliers of management or equipment. It was suggested that the financial institutions should seek ways to liberalise terms for projects that contribute to the national benefit. In fact this is often the case with such projects as highways where, in any case, rates of return are difficult to estimate. The relation between the extent of debt financing and project autonomy was pointed out. Accurate cash-flow analysis is essential to ensure debt service. The unreliability of market predictions suggests that only part, though perhaps the major part, of the product be sold on long-term contracts and the rest on the spot market. The favourability or otherwise of payment for machinery supplies partly in cash and partly in kind was discussed. A difficulty is the commodity price fixed for the "in kind" payment and there are dangers that high prices will be charged where the supplier has the power to fix prices. On the other hand an advantage is the speed with which the financing can be arranged.

Risk analysis, an understanding as to how to proceed in a joint venture when one party fails to deliver promised results and how to relate cash flow statements to the valuation of a mine in a situation with changing prices, are essential elements in financial management of mining enterprises. Not least in importance is the decision whether to make use of new and rapidly improving technology when finance considerations favour the use of tried and proven methods. It was proposed as essential that new technology must promise high returns; the greater the returns the greater the degree of risk-taking justified in this respect.

CLOSING ADDRESS TO THE SEMINAR

by Geoffrey Robson

As this meeting is drawing to a close I would like to make a few concluding remarks which are not intended to summarise the talks by our Lecturers or our discussions, but rather to highlight some of the points that have been made by both Lecturers and Participants.

During this meeting on the Economics of Mineral Engineering we have been concerned with two rather different aspects of mineral resources development—on the one hand, with mineral processing and metallurgical technologies as they relate to the extraction of metals and concentrates from ores; and on the other, with the social, economic and environmental issues as they relate to mineral development.

The view has been expressed in this meeting, and generally accepted, that mineral resources should be used not simply as assets for commercial exploitation but as a basis for social and economic development through education and the development of skilled manpower, communications and administrative capabilities. If then, mineral development is to be the base for economic and social development in developing countries, how is mineral development to be achieved? Our discussions have highlighted the fact that at the heart of many of the problems of mineral development in developing countries lies the question of finance. In many cases finance for mines and processing plants is obtained—through various agencies—on the international money markets. In this case, as we have heard, the lenders are concerned that their investment will be returned in full together with interest at a rate appropriate to the risks involved and the investment alternatives and will therefore wish to see that their risks are covered, such as by long-term production contracts, and to see that those involved have experience in the production, processing and refining of metals and their ores.

We have heard of the dangers of too rapid progress from mining into processing; that a smelting plant should not be linked to a single mine; that an integrated mining and processing project may be difficult to finance, and that projects involving the most modern technology should not be carried out first in developing countries. A participant at this seminar from a developing country might be forgiven for thinking that our lecturers, for the most part from developed countries, seem to be discouraging mining development except on their own terms and to be taking too negative an attitude to the possibilities in the developing countries.

Personally, I do not think that our lecturers have this attitude; I believe their cautionary remarks and their attention to the details of the many possible problems that may arise in mining projects stems from the fact that when a large mining project in a developing country relies on the international money market for its finance it competes with other similar projects in developed countries and therefore is subject to the same critical scrutiny as all other similar projects of both the developed and the developing countries.

Those mineral developments that are proposed by developing countries with resources of foreign exchange to invest may not experience the same difficulties in launching their minerals projects, either because they provide all the finance required themselves or because they provide appropriate guarantees to the external investors. It has been emphasised that the development of mining, processing and refining on a scale suitable for the supply of local manufacturing industry is a worthwhile objective for developing countries and, in many cases, can be achieved with modest local reserves of foreign exchange. Where mines and processing plants are developed for the export market using domestic foreign exchange resources, the developing country will have to consider what effect its new production will have on product prices and also whether its national social and economic goals will be better served by alternative investments in other, possibly more rewarding, sectors of its economy.

Much of our seminar has been devoted to discussions of mineral processing and refining technology, and we have heard how factors such as the requirements of pollution control and other environmental questions, the rise in energy and other costs and the need to process more difficult ores are leading to new technologies and in particular to a trend towards the use of hydrometallurgical techniques in refining. We have also heard how hydrometallurgy is being developed as a mining technique by *in situ* leaching of ores. We have discussed environmental issues and the penalties in higher costs that are incurred when pollution controls are in effect, but the importance of environmental controls as the scale of mining increases is apparent to all of us.

In concluding, I would like to thank you all, lecturers, participants and observers, for your interest in this meeting and for your very valuable contributions to it.

The very considerable organisational effort made by our hosts on our behalf and the kindness and unfailing courtesy with which they have responded to our needs have been a large factor in determining the success of this meeting and also our pleasure in participating in it. I am sure I speak for all of you when I express to Dr. Alpan and his colleagues at MTA our warm appreciation and thanks. I would also like to thank Mr. Shallon and the staff of the United Nations office in Ankara who have laboured hard on our behalf, and, finally, to thank our interpreters who have very competently managed our technical discussions and assisted so well in the smooth running of our meeting.

STATEMENT BY THE PARTICIPANTS

WE, THE PARTICIPANTS TO the Interregional Seminar on Economics of Mineral Engineering held at Ankara, Turkey, from 5 to 15 April 1976, whose names and addresses appear on the attached list, wish to put on record our sincere appreciation to the United Nations and the Government of Turkey for organizing the Seminar, and hereby submit the following statements to the United Nations for its consideration:

1. That our discussions at the Seminar have drawn attention to the financial aspects of mineral resources development as an issue which is of special concern to the developing countries;

2. And therefore ask the United Nations to request both international and regional financing institutions to make funds available for mineral resources development in developing countries;

3. That the participants have realised the need for mutual co-operation amongst their various countries particularly in technology transfer for mineral resources development, investments and marketing;

4. The participants noting recent cuts in UNDP assistance in the field of mineral development in developing countries emphasize the need for the United Nations to send experts to the developing countries, when requested, to conduct an objective and impartial assessment of mineral resources development projects and offer the necessary recommendations;

5. That the United Nations may collect and publish information on mineral resources development experience in the fields of mining engineering, mineral processing and metallurgy from various developing countries which already possess the necessary know how and technology for the benefit of those concerned;

6. That the United Nations may regionalise training facilities for the benefit of developing countries insofar as mineral resources development is concerned;

7. That the proceedings of this Seminar be published as soon as possible;

8. That finally, due to limited time available for discussion of financial aspects of mineral resources development during the Seminar, the United Nations is requested to organize another Interregional Seminar on financing and mineral economics with special reference to developing countries as soon as possible.

LIST OF PARTICIPANTS

BOLIVIA	**Alvaro Guillermo Ugaldo** Director of Planning Ministario de Mineria, Metallurgia La Paz
BRAZIL	**Frederico L. M. Barboza** Chief of Economic Section Department of Mines (D.N.P.M.) Ministerio de Minas e Energia RECIFE
BURMA	**U. Sein Hla Aung** General Manager Myanma Bawdwin Corporation Rangoon
BURUNDI	**Zacharie Kambirigi** Directeur des Mines et de la Géologie Ministère de la Economie Bujumbura
CHILE	**Carlos Molina** Head Extractive Metallurgy and Chemical Industry Departments, INTEC Santiago
EGYPT	**Abdalla Wassef** Geologist Geological Survey of Egypt Cairo
GHANA	**Godfried Opong Kesse** Director Geological Survey Accra
INDONESIA	**Charis Chusno** Head, Planning and Programming Division Ministry of Mines Jakarta
IRAN	**Ali Ghandiri-Asli** Expert responsable Minière Ministère de l'Industrie et des Mines Teheran
ISRAEL	**Michael Beyth** Co-ordinator of Earth Science Ministry of Industry and Commerce Jerusalem
IVORY COAST	**Joseph N'Zi** Directeur Général Société Pour le Developpement Minière De la Côté D'Ivoire (SODEMI) Abidjan
KENYA	**C.Y. Ochieng Owayo** Chief Mining Engineer Mines and Geology Department Ministry of Natural Resources Nairobi

KOREA	**Mahn Sup Kim** Section Chief, Planning Section Geological and Mineral Institute of Korea Seoul
MADAGASCAR	**Solobert Jacky Ramaroson** Direction de l'Industrie et des Mines Tananarive
MALI	**Mory Kane** Chef, Service Prospection Géologique et Minière Direction Nationale de la Géologie et des Mines Bauyako
MEXICO	**Sergio Sosa** Gerente Produccion Minas Consorcio Minero Peña Colorado Manzanillo
MOROCCO	**El Mouloudi Arsalane** Chef de Valorisation des Minerais Ministère du Commerce, de l'Industrie des Mines et de la Marine Marchande Rabat
NEPAL	**Mahendra N. Rana** Director Bureau of Mines Ministry of Industry and Commerce Lainchaur, Kathmandu
PHILIPPINES	**Victor de los Santos** Mines Regional Officer Bureau of Mines Manila
RWANDA	**Pierre Claver Mahungiro** Chef de Division Production Minière et Inspection Ministère des Ressources Naturelles Kigali
THAILAND	**Kreetha Jitrapatima** Director, Mining Technology Division Department of Mineral Resources Bangkok
TUNISIA	**Foued Lakhoua** Directeur Inspecteur General Montfleury
TURKEY	**Dr. Sadrettin Alpan** General Director Mining and Exploration Institute of Turkey (MTA) Ankara
	Dr. Altan Dogu MTA Institute Ankara
	Prof. Zeki Doğan Middle East Technical University Ankara
UPPER VOLTA	**Pierre J. E. Tapsoba** Directeur Direction de la Géologie et des Mines Ouagadougou
YUGOSLAVIA	**Mrs. Zora Djukie** Dipl. Engineer—Metallurgy Institute for Technology of Nuclear and other Raw Materials Belgrade

ZAIRE	**Biterwa Barore** Bureau d'Etudes Department des Mines Kinshasha
	[Observers]
E. Machens	Organization for Economic Co-operation and Development (OECD) Paris, France
R. Lalkaka	United Nations Industrial Development Organization (UNIDO) Ankara, Turkey
M. Buckingham	United Nations Industrial Development Organization (UNIDO) Ankara, Turkey
	[Lecturers]
Michael West	Editorial Director, The Mining Journal Limited, London, England
Henry E. Cohen	Professor of Mineral Technology, Department of Mining and Mineral Technology, Royal School of Mines, London, England
Douglas N. Moir	Warren Spring Laboratory, Stevenage, Hertfordshire, England
T. R. Scott	Chief Research Scientist, Division of Mineral Chemistry, CSIRO, Australia
Michel Turpin	Director of Research, Ecole Nationale Supérieure des Mines, Paris, France
Roshan B. Bhappu	Vice President, Mountain States Research and Development, Tucson, Arizona, USA
B. R. Nijhawan	Senior Interregional Adviser, UNIDO, Vienna, Austria
John H. Harris	Technical Adviser, Minerals Section, Centre for Natural Resources, Energy and Transport, United Nations, New York, USA
Josef Lajzerowicz	Department of Energy, Mines and Resources, Ottawa, Canada
Peter A. Glenshaw	Consultant, Reston, Virginia, USA
	[Secretariat of the Seminar]
Director	**Geoffrey R. Robson,** Centre for Natural Resources, Energy and Transport, United Nations
Co-Director	**Sadrettin Alpan,** Mineral Research and Exploration Institute, Turkey
Rapporteur	**John H. Harris,** Centre for Natural Resources, Energy and Transport, United Nations
Technical Secretary	**David Dreiblatt,** Centre for Natural Resources, Energy and Transport, United Nations
Administrative Secretary	**John Ditchburn,** Office of Technical Co-operation, United Nations